Praise for The 13th Demon: /

"A gripping look into the
will keep you turning pages –
last."

– Mike Yorkey, co-author of Chasing Mona Lisa and the Every Man's Battle series

"Bruce Hennigan's The 13th Demon, Altar of the Spiral Eye is a thrill-a-second ride and an impressive debut. Intense, gripping, suspenseful, and spooky – once you open the cover, the pages will fly. Carve out some time for this one … you're gonna need it."

– Mike Dellosso, author of Frantic, Darkness Follows, and Scream

Praise for The 12th Demon: Mark of the Wolf Dragon:

"Bruce Hennigan has done it again. The 12th Demon is everything a good suspense novel should be and will have you realizing you've been holding your breath for way too long. Interesting characters, a plot that speeds along, and a message you can ponder for a very long time convince me of two things … The 12th Demon is a hit, and Bruce Hennigan is the real deal."

– Mike Dellosso, Author of Frantic, Darkness Follows, and Scream

"Jonathan Steel fans rejoice! Bruce Hennigan is back with a quick-paced thriller chock-full of action, plot twists, and some great slam-bang monster action. The 12th Demon is a fun read that puts a fascinating new spin on the vampire genre."

– Greg Mitchell, author of The Strange Man and Enemies of the Cross

"Bruce Hennigan's The 12th Demon is the redemptive answer to books like the Twilight series."

– Mike Yorkey, co-author of Chasing Mona Lisa and the Every Man's Battle series

Also by Bruce Hennigan:

The Chronicles of Jonathan Steel
The 13th Demon: Altar of the Spiral Eye
The 12th Demon: Mark of the Wolf Dragon
The 11th Demon: The Ark of Chaos

The Conquest Series
Conquering Depression: A 30-Day Plan to Finding Happiness
Hope Again: A 30 Day Plan for Conquering Depression

The 10th DEMON

Children of the Bloodstone

By Bruce Hennigan

ar813ea

Copyright © 2015 Bruce Hennigan

All rights reserved. No part of this book may be used or reproduced by any means, graphic, electronic, or mechanical, including photocopying, recording, taping or by any information storage retrieval system without the written permission of the publisher except in the case of brief quotations embodied in critical articles and reviews.

The 10th Demon: Children of the Bloodstone
By Bruce Hennigan
Published by Area613
An imprint of 613media,LLC
10911 Sanctuary
Shreveport, LA 71106
www.steelchronicles.com

Cover and layout design be Fusiform Design Workshop, llc

This is a work of fiction. The characters portrayed in this book are fictitious unless they are historical figures explicitly named. Otherwise, any resemblance to actual people, whether living or dead, is coincidental.

Author's website www.brucehennigan.com

All scripture quotes are from the New King James Version of the Bible.

Demographic info: ISBN, Library of Congress notation, etc.

Author's Note:

In "The 12th Demon: The Mark of the Wolf Dragon" the reader met Dr. Renee Miller's brother. In my original manuscripts, this person was Dr. Miller's sister. After having to meet strict word count restraints, I had to rewrite that pivotal scene in my second book. In this book, I am restoring the character to her original identity, the sister of Renee Miller, Dr. Cassandra Sebastian, also known as "The Artifact Hunter".

"Does science inadvertently – and dangerously – open Pandora's box in affirming that nature and supernature interface? The fear that it does drives many scientists and others away from any willingness to discuss religious beliefs in general and paranormal in particular. Meanwhile, humanity proves persistently religious and incurably curious about the paranormal and supernatural."

Dr. Hugh Ross in "Lights in the Sky and Little Green Men"

The pride of your heart has deceived you,
You who dwell in the clefts of the rock,
Whose habitation is high;
You who say in your heart, 'Who will bring me down to the ground?'
Though you ascend as high as the eagle,
And though you set your nest among the stars,
From there I will bring you down," says the LORD.

Obadiah 1:4-5 (NKJV)

PART I

DISTANT SIGHTINGS

Nocturnal Lights (NL): Phenomena observed at night, particularly unconventional lights.

Daylight Disks (DD): Phenomena observed during the day, usually oval or disk shaped objects in the sky.

Radar-visual (RV): Unexplained radar blips that coincide with visual sightings of UFOs.

J. Allen Hynek's category of UFO Phenomena
Center for UFO Studies

Chapter 1

Close Encounters of the third kind (CE-3): *The living occupants of the craft (UFO) are seen by witnesses; however, no communication or further contact takes place. This kind of encounter almost always happens at night and usually with only one or two observers.*

J. Allen Hynek

Quantum flux surveillance:
30.681768,-86.949234
2:46 A.M. CDT
Grimvox Raptor Neural Interface
Species "Homo sapien sapien"
Subspecies Agkistrodon piscivorus;
"Water Mocassin"

The receding clouds of hurricane Leo swirled into a scorpion shape along the southern horizon. From within the whirling mists, the constellation Scorpio burned through and shone brightly in the man's night vision goggles. He squatted on the stump of a pine tree snapped off by the winds and the fragrance of pinesap did nothing to dispel the fetid odor of the swamp. He touched a control on the side of his goggles and the heads up display from his drone popped into view. The drone hovered just above him in the still night and he twitched his right eye to send the drone upward through a haze of hungry mosquitoes.

Hurricane Leo had decimated the swamp. Hardwoods

once draped with Spanish moss and tall pines had been leveled by the destructive winds. In the aftermath of the storm, silence haunted the air and tons of silt and brackish water stifled the normal orchestra of swamp animals and insects. The man felt the water moccasin before he saw it and without taking his eyes off of the heads up display, he kicked it away with one quick movement. The snake somersaulted into the darkness and splashed into the rancid waters.

The view from the drone spun away from Scorpio and he had no problem surveying the logging road cutting through the leveled forest. He was certain the girl would come down that road. Unlike the rest of her kind who would be huddled in fear in the basement of their school, she would run and try to find the Guardian. He would be waiting for her and this time, those who wanted to put her under the dissection knife would not find her. He had protected her for six years and a hurricane wouldn't stop him now.

The drone moved above the logging road and a figure appeared in infrared. Her heat stamp was as unmistakable as her remarkable speed. The figure paused and he watched the hot pinpoints of her eyes turn upward. She had spied the drone and one arm swooped toward the ground. She hurled the rock upward and he barely maneuvered the drone to dodge it. No normal human would have been able to throw that high. He recalled the drone and told it to hover above him.

A tall figure of power and speed appeared through the broken trees. She stopped a meter away and even though he squatted four feet above the ground she still towered over him. In the time since he had left her with the others to journey to Louisiana, she had grown a foot. At the age of eight, she was no ordinary girl.

"Who's there?" She whispered.

"The Guardian." The man said.

Brown water had stained her one-piece jumpsuit and her blonde hair was pulled into a ponytail behind her head. "Beneath the Patagonian giant." She said.

"The bloodstone glows with life." The Guardian said.

"My Guardian!" She smiled and pointed back over her shoulder. "The storm knocked out the school's perimeter alarms."

"Where are the others?"

"Hiding in the tunnel beneath the school. They are so silly! I knew I was safer with you."

"You saw my drone?"

"Yeah! You have a new toy. I'd like to fly--"

His night vision goggles fell into darkness as they inexplicably lost power leaving him blind. He slid the goggles from his eyes and looked up at the drone. It tumbled from the sky and landed in the water next to the stump. Above them something moved against the dark night. In total silence it blotted out the bright stars.

"What is it?" She whispered.

"I don't know but we need to go. Now! East toward my motorcycle."

"I'll carry you." She snared him by the waist and threw him over her shoulder. He gasped for air as she moved with incredible speed and deftness. She needed no goggles with her superior night vision. A cone of purplish light erupted from the sky and focused on the downed drone. The cone swiveled toward them. He reached into his backpack and pulled out a flare, ignited it and hurled it to the north of their position. The light seemed to displace itself in seconds to hover over the flare.

They came to a sudden halt and she placed him on the ground. The sound of rushing water filled the air and he looked down a steep bluff into the churning waters of a river gorged on rainwater. "No! We need to be further north!"

The air filled with a sickening fragrance like fermented honey and rotten fruit. An exotic, unearthly music underscored the fragrance. A low, throbbing base sound wormed its way into his skull. He clamped his hands over his ears but the music penetrated to his core. He whirled and the cone of light appeared above the girl. It encased her in purplish light and he grabbed her arm but she was paralyzed.

"We have to go!" He shouted above the music. Above

them the reflected light illuminated the smooth, metal swoop of a disc and in its center, an iris opened emitting lurid green light. Something skittered out onto the surface, all dark twittering wings and spindly limbs. It jerked and gyrated on the undersurface of the craft. Cold gripped his legs and the numbness moved up his body. The girl hovered upward into the air toward the winged creature and the pale green craft. The man's arms would not move. Soon, he would join her. Unless!

"I will find you, Vega! I will find you!" The Guardian shouted and teetered backward. He slid head first on the muddy ground and fell end over end over the bluff. He held his breath, the only control he had left and hit the water hard. He plunged deep into the turbulent current and spun end over end until his pack snagged a log on the bottom of the river. His night vision goggles slid down over his face and sprang to life. The drone was alive again and following its programming, it would return to his location. Feeling crept back into his arms and he finally managed to slam his numb fist against the backpack release. He slid out of the arm straps and stroked toward the surface. His head broke into cold air and he sucked in welcome air.

With his goggles activated he watched the drone move through the night toward the craft. From above he made out a perfect disc with a sloping surface and a dome in the center. The drone swooped beneath the craft and he watched Vega disappear into the iris. The creature on the craft launched itself into the air. Its four wings vibrated as it moved toward the drone and its unearthly face filled the camera. Its head blurred and spun, blurred and spun in a nightmarish carousel of white-rimmed eyes of deep black and a toothless maw filled with red fire.

The Guardian ripped away the goggles. A floating log rammed him in the back. He grabbed it and pulled himself up against the current. He glanced over his shoulder. The saucer shaped craft hovered against the night sky and with a sudden motion it disappeared leaving behind a purple streak in the night.

Chapter 2

"Animals, especially dogs and cattle, but also rabbits, goats, and horses, have shown noticeable agitation in the presence of UFO phenomenon."

Hugh Ross, "Lights in the Sky & Little Green Men"

```
Quantum flux surveillance:
30.495352,-87.251186
7:46 A.M. CDT
Grimvox Mammalian Neural Interface
Species "Canis lupus familiaris"
```

"So, this lady says she's lost her 'Poopems'?" Theo King asked.

Jonathan Steel stood on the debris littered lawn of the small house. Considering a hurricane had passed within miles of the area, little damage had been done. In contrast to the tiny gnome guarding the flowers around the front porch, Theophilus Nosmo King towered over six and one half feet tall and tipped the scales at three hundred fifty pounds. He wore a black and gold Saints football jersey and long athletic pants. He pulled off his sunglasses and wiped sweat from his bare head and from his face. "Jonathan, who would call their husband Poopems?"

Steel put a piece of gum in his mouth and looked up at his partner. "We are what we do."

"Ah, I get it. He's like Cephas. The old man shouldn't eat cabbage."

"He likes cabbage."

"Yeah, but cabbage don't like him! What does losing 'Poopems' have to do with demons?"

The sun burned with incredible intensity after the clouds had receded. Already, sweat soaked Steel's tee shirt and dripped down his legs inside his jeans. "I don't know. I'm just doing my next-door neighbor a favor." Steel said.

Theo raised an eyebrow and the sunglasses sitting on his forehead moved with it. "You have a neighbor? Someone came over to your house and had the courage to knock on your back door and ask for coffee or something?"

Steel gave him his best acid look. "Are you saying I'm not the friendly sort?"

Theo shrugged and pushed his sunglasses back down over his eyes. "You just as soon bust somebody's chops as say 'good morning'. That's all I'm saying."

Steel took off his cap and wiped sweat from his short hair. "There's a lady lives in the condo next to mine. I asked her husband to hurricane proof the condo while I was gone. So, I owe her a favor."

"Yeah, you don't like owing people favors. And, she probably all wobbly from those eyes of yours." Theo said.

Some people found Steel's bright turquoise eyes beguiling. He found them a nuisance. "She was probably color blind, Theo."

"Fine with me. Not a whole heap of damage here, eh Chief?"

There was that name again. He had never liked Theo calling him Chief. Until he learned it was a name of endearment used by Theo's late grandmother. It signified respect. Steel looked down the street lined with oak trees and palm trees. Hurricane Leo had only skirted this Pensacola neighborhood. "Doesn't look like it."

Theo nodded and glanced up at the burning sun. "Hot as Hades out here. When we get inside, I'll turn up the AC. You talk to the little old lady. Let's go find Mr. Poopems."

Theo walked across the lawn toward the small house. Just a few weeks before, Theophilus Nosmo King had been a drugged up homeless wreck of a man. Steel had seen something worth saving in the man when Theo had tried to rob him one night outside a shopping mall. He didn't know why he sometimes heard the still, small voice inside telling him to "save" someone but he was glad he had heeded it that night. Theo had helped save Josh Knight from the twelfth demon, Rudolph Wulf. And, he had been by Steel's side when they faced down the eleventh demon. Since that time, Theo had sobered up and was now his partner in this odd business of tracking down evil in the lives of ordinary people.

Steel took off his sunglasses and wiped sweat from his face. For a second, he studied his reflection in the mirrored sunglasses. His bright, turquoise eyes gleamed in the sunlight and once again he wondered who he was. For over a year now, he had tried to recover the lost memories of a lifetime as he pursued the powers of darkness across the land. So far, only a few painful and confusing memories had returned. He sighed and pushed the sunglasses back over his eyes. He spit the gum into the flower garden. It never lasted very long. Too much trouble.

"You coming, Chief or are you gonna get some more sun?" Theo asked from the front porch.

"And then Elvira just floated up into the air right up next to the light." The old woman gestured toward the chandelier above the dining room table.

"Mrs. McGilacutty, Elvira floated up past the chandelier and then what?" Steel asked with a straight face.

"Elvira just disappeared. Right through the ceiling."

"Did your husband go and try to find Elvira?"

"Heavens no!"

"Then where did your husband go?" Steel said

"Henry has been dead for four years." Mrs. McGilacutty said.

"Mrs. Allen asked me to come help you. I thought your husband was Poopems."

"Elvira is my Poopems, young man." She dabbed at her powdered nose with a tissue. "Mrs. Allen said you investigated the paralyzed normal."

"Paranormal." Steel gritted his teeth. Be patient, he told himself. Don't let that anger get the best of you. She's just a helpless old lady. He glanced at Theo as he appeared from the hallway. This is what he got for making promises. "Mrs. McGilacutty, I don't investigate the paranormal. I just help people, well, with evil."

"And what would you call something that can suck my Poopems right through the ceiling, eh?"

"She got a point there, Chief." Theo towered over Mrs. McGilacutty. "I looked all around the house and I didn't see any signs of forced entry."

"We're looking for a dog."

Theo lifted his right foot and studied the underside of his running shoe. "That explains the smell. You are what you do."

"Poopems is more than a dog!" Mrs. McGillacutty stood up and pointed the tissue at Steel. "She is my friend. Now, Esther told me you were a good man who would help me. So, now that you're here, find my Poopems. She's been gone for three days. And, the things she said to me!"

"She talked to you?"

"Yes. She's never said anything to me. Of course, sometimes I talk for her in a little bitty voice. I know what she is thinking, you know." Mrs. McGillacutty said. "That's how I knew something pair of normal was going on because she would never have talked that way to me."

"What did she say?"

Mrs. McGilacutty swallowed and pressed the tissue against her face. "No, don't make me say it."

Steel reached over and pulled the woman's hands from her face. "Mrs. McGilacutty, look at me."

Her eyes widened. "You have such gorgeous eyes. Like chips of turquoise. Henry had a necklace with a piece of turquoise in the shape of a turtle." She nodded and pressed the tissue against her right temple. "My head hurts. I need my Poopems."

"What did your dog say to you?" Steel said.

"Fowl things! Nasty things! X-rated!" She spat the words. "She broke my heart. No one has ever said such things in my house. No one." She paused and drew a deep breath and glared at Steel with her red-rimmed eyes. "And then, Mr. Steel, she said something odd. 'The Children of Anak are coming'." Mrs. McGilacutty drew a deep breath. "Did they come and take my Poopems?"

"Who is coming?" Theo asked.

"The Children of Anak." Mrs. McGilacutty nodded. "Yes, that was it. A. N. A. K. She even spelled it for me."

"Your dog can spell?" Theo asked.

"I told Henry she was gifted." Mrs. McGilacutty nodded and let forth a tiny giggle. "He said she was just a dog. Well, he's gone and Poopems is still here --" She jerked and reached out and grabbed Steel's hand. "You have to find her. Mr. Steel, my Poopems told me to call Esther Allen and get her to get in touch with you. She said only you could find her."

Under other circumstances, these facts would have proven the woman's senility. But, Jonathan Steel's experiences of the past few months had opened the door onto a supernatural world where everything Mrs. McGillacutty described was not only possible, but rational. Steel stood up and studied the beige ceiling. "I guess you aren't crazy after all."

"What?" Mrs. McGillacutty said.

"Anak?" Steel glanced at Theo. "Recognize the name?"

Theo's forehead wrinkled with thought. "Maybe in the Old Testament. I'd have to look it up." Before Theo had gotten into drugs, he had been a preacher at a church in California. Among other things.

"It's there, all right. I can't put my finger on it but it's there. You think little Poopems had a visitor?" Theo said. He put his thick hands in the air and waved them around in a ghostly fashion. "Say, maybe, Satan!"

"What?" Mrs. McGillacutty glanced at Theo.

"Just calm down, Mrs. McGillacutty." Steel made the

effort to pat her bony shoulder. "Mr. King is just kidding." He glared over his shoulder at Theo and the man shrugged his huge shoulders and pointed to the ceiling.

"I think you had best be climbing on up into the attic, Chief." Theo whispered. "It's hot up there and I'm sure old Lucifer is right at home." He grinned and pushed his sunglasses back down over his eyes. "I'll stay here and protect Mrs. McGilacutty."

"Very funny." Steel looked down the hallway at a rope hanging from a closed doorway in the ceiling. "Is that the ladder that leads up into the attic?"

"Yes it is. Now you be sweet to Elvira." Mrs. McGilacutty said. "And, if she talks to you, you answer her back. She has a sensitive personality."

Steel stepped off the top rung of the ladder into smothering heat. Hot air hovered in the attic from the simmering sun just on the other side of the roof. A year ago, he would have laughed at Mrs. McGillacutty's claim that her dog had spoken to her. But, that was before he had met Rocky Braxton and his mentor, the thirteenth demon. Steel had been pulled headlong into the battle between the forces of good and evil and had found himself a reluctant draftee. He had seen things in the past year that would drive most men mad. The idea of a speaking dog telling its owner about the "children of Anak" was just another minor incident in a typical day.

He pulled a small flashlight from his back pocket and its tiny cone illuminated dust particles dancing on the heat waves. Boxes piled haphazardly around the attic threatened to ignite in spontaneous combustion in the heat. The odor of old cardboard and dry dust filled his nostrils. He wiped sweat from his face and studied the layout of the ceiling joists to determine the location of the dining room. Steel stepped from wooden beam to wooden beam avoiding the insulation until he rounded a vertical support and made out the peaked roof over the dining room. Above him, the tiny furred body of a dog slowly rotated like some errant planet.

Steel had seen human sacrifices. He had been attacked by a giant scorpion. He had fought off an army of vampires and had been attacked by a white-eyed ghoul. But, the sight of something so ordinary gripped in the claws of the supernatural filled him with a primal terror. He drew a deep breath and said a quiet prayer for help.

He stepped slowly across the beams until he stood just beneath the rotating body of Elvira. The dog seemed paralyzed and no worse for wear for having hovered in the stifling heat of the attic for the past three days. When his flashlight beam hit the dog she came to life. Elvira began to wiggle as if struggling against some unseen force until she looked down at Steel. Intelligence filled the dog's eyes and a tiny voice issued from her snout.

"Ah, Jonathan Steel! We've been waiting for you. It's time for us to play. The sons and daughters of Anak are here." Mist poured from the dog's mouth. Steel shuddered as the heat receded in the wave of moist, frigid air.

"Who are you?" His breath steamed.

"Friends of your adversary. And he will see that you are swallowed up by the mouth of Satan! Swallowed whole by Satan down his gullet, down his mouth!" Elvira said. She fell onto his head and he stepped backward onto soft, mushy insulation. The sheetrock ceiling cracked beneath his weight and he tumbled backward. He thudded against wooden beams as he plummeted and he grabbed Elvira in a desperate hug. He landed on Mrs. McGillacutty's dining room table. It cracked beneath his weight and he came to rest on the floor surrounded by insulation, sheetrock, and broken table bits. Mrs. McGillacutty smiled as she pulled Elvira from his grasp.

"You found my Poopems."

Chapter 3

". . . the absence of physical proof, indeed even of solid circumstantial evidence, indicates to the UFO conspiracy theorist just how secret and manipulative the U.S. government is when it comes to alien visitation."

Mark Clark, "Lights in the Sky & Little Green Men"

"Ouch!" Steel jerked as the needle pierced his forehead. Dr. Garcia rolled his eyes above his surgeon's mask.

"Hold still! I'm on the last stitch."

He relaxed back onto the gurney. The emergency room doctor snipped at the stitches with a small pair of scissors. "What were you doing in the attic? Looking for antiques?"

"No, a dog."

Dr. Garcia paused and his eyes reflected a careful measure of concern. "A dog? In the attic?"

"It's a long story."

Garcia held up two fingers. "How many fingers?"

"Two. Mrs. McGillacutty lost her dog in the attic."

The skin around Garcia's eyes wrinkled and he chuckled. "The old lady in the waiting room with the big man?"

"Yes. How much damage did I do?"

"Probable concussion, three lacerations on your scalp and back, and a broken rib."

"I meant at the house."

Garcia regarded Steel with his impenetrable gaze. "I wouldn't know. But, judging from all your scrapes and scratches, I'd say the damage is considerable. I hope the old lady has good insurance."

A woman in scrubs appeared over Garcia's shoulder. "Dr. Garcia, there's a soldier demanding to see one of the other patients."

"Soldier?" Garcia pulled the mask away from his face.

"A Major somebody. He demands that you come speak to him immediately."

"If you'll excuse me, Mr. Steel, I need to kick some brass." Dr. Garcia left the cubicle. Steel sat up and pain lanced across his left side. He drew a deep breath and ran his hand under his tee shirt and massaged the painful rib. Angry voices came from the other side of the curtain.

"Sir, I don't care if you're the president of the United States. I am in charge of this emergency room and you will step out into the waiting room. I will see you as soon as I finish with my patients." Garcia said.

Metal grated on metal and an unmistakable click filled the air. "Do you see this gun, Doctor? It gives me the authority to do whatever I want. This is a matter of national security and if you want to be able to continue to take care of your patients, you will step out of the way."

The curtains from the adjacent cubicle stirred and a man appeared. His scratched face was bloody and his clothes were torn. Streaks of gray and red mud ran through his long, black hair. Dirt and dried blood caked his beard. His mouth fell open.

"Jonathan Steel?" He whispered.

The man's features were familiar but before Steel could think any further, the man slid onto the shelf beneath Steel's gurney. "If you tell them I am hiding, it will endanger Josh Knight's life." The sheet fell back into place and hid the man's face.

A soldier dressed in desert camouflage stepped into the cubicle. His hair was shorter than the last time Steel had seen him in the picture on Renee Miller's shelf. Renee owned Ingenetics Laboratories in Dallas, Texas. She had helped him defeat the twelfth demon and rescue Josh. Her ex-husband was the man who stood before him. Major Robert Miller's dark eyes registered a flicker of recognition before his intense gaze moved around the room. "Did a man come in here from the cubicle next to you?"

"No. Who is he?"

"You don't need to know." Miller threw aside the curtain to the next cubicle. "But, he was involved in an incident that impacts national security."

"I heard someone in the other bed earlier." Steel said. "I know you. I saw your photograph on a shelf in your wife's office in Dallas."

Miller froze and he glared at Steel. "My ex-wife, you mean. How do you know Renee?"

"Professional association." Steel said. "She supplied me with an aircraft."

Miller stiffened. "Those were once my aircraft." He grabbed a radio microphone clipped to his collar. "Secure the perimeter of this hospital. I know he's here."

The sound of scuffling feet and the shadows of rushing bodies passed on the other side of the curtain and Miller was gone. Steel looked beneath his gurney. Empty. Was the man from Steel's past? With a lifetime of lost memories to sift through, he could be anyone. He stepped into the other cubicle. Dirt and rock chips littered the bed. Brown water stained the sheets.

"Mr. Steel, there you are." Garcia said as he pulled aside the curtain. "Where's John Doe? They were looking for him! I knew there was something strange about the man."

"What happened to him?"

"He fell down a bluff into Blackwater River over towards Bagdad."

"Iraq?"

"Bagdad, Florida. It's just north of Pensacola. Good thing the river was high after the hurricane or the fall would have killed him. He said he drifted on a log out into Pensacola Bay." Garcia regained his professional composure. "Patient confidentiality, Mr. Steel. Now, back to your bed."

Steel stepped back into his cubicle and sat on the edge of

the gurney. Who was this man? And why had he mentioned Josh's name? Major Miller, Renee's ex husband surfacing in the midst of all of this? What was happening? The pain medicine muddled Steel's thoughts. But, he had to figure out the identity of the strange man. Josh's life might depend on it.

Chapter 4

"Let's face it. Life on Earth stinks! That is why we look to the heavens for our answers to all of life's painful crises. I've discovered there are as many answers to life's questions buried in the ground. Those are the secrets I uncover on this television show."

Dr. Cassandra Sebastian, The Artifact Hunter

"And, then the spear pierced my skin, just below my pecs." Josh Knight tried his best to contract the muscles underneath the two crimson scars on his chest.

"I am sure there is a logical reason why you are talking to your chest." Cephas Lawrence said.

Josh spun away from the beach. "I was just, uh, practicing."

Cephas looked down at the sand stretching away from the deck of Jonathan Steel's beach house. It amazed him how quickly people returned to the beach after a hurricane. Near the water's edge, two teenage girls threw a flying disc. "Ah, you were practicing recounting your near death at the hands of Rudolph Wulf, the twelfth demon so as to impress those two young women."

Josh massaged his scars. "Yeah, that's about right, Uncle Cephas."

Cephas held up two plastic zip bags filled with ice. He pressed them against Josh's bare chest. Josh winced and lurched away.

"Dude! What's up with the ice?"

"Originally, they were for your hands." Cephas wiped the condensation from his hands on his tee shirt. He wore baggy shorts below a shirt bearing the huge smiling face of Albert Einstein sticking out his tongue.

Josh held the dripping bags away from him. "There's nothing wrong with my hands. And, what's with the tee shirt. He looks like your twin. And, that pink sunscreen on you nose? Dude, really?"

Cephas massaged his bushy moustache and shrugged. "I deduced you must have had something wrong with your hands since you can't seem to tear yourself away from watching those two young girls frolic on the beach and help me finish packing. Perhaps the ice will bring down the swelling."

Josh frowned and cast one last look at the two girls. "Uncle Cephas, I just needed a short break. I mean, we're here on the Gulf Coast in August and the babes are really hot and you have me inside packing up dusty books and stuff."

"Jonathan said we needed to have all the boxes in the den packed by the time he got back." Cephas said. "And, you've been out here for over an hour. In fact, you might want to put your shirt back on. Not only do you lack the pectorals that would attract those girls' attentions, but you are quite sunburned."

Josh glanced down at his chest. "Ouch." He pressed the ice packs against his skin and sighed. "Ah, that's better."

"Indeed. Now, if you will come inside you can help me finish the packing." Cephas pointed over his shoulder toward the huge beach house. He glanced up at the sun. "It's a scorcher out here and if you don't help me, it will be a tad bit hotter inside when Jonathan returns and the boxes are not packed. You know how angry he can get."

"Dude, there are constants in the universe. The speed of light, the value of pi, and Jonathan's temper." He sighed and shrugged. "I guess these fine ladies will never know what they could have done with the Joshman."

"The Joshman?"

"Okay, so it's pretty lame."

"Come inside before you embarrass yourself." Cephas said.

"You have a lot of room to talk, dude." Josh pointed at his legs. "I've seen flamingos with more muscular legs. Especially with the pink nose." Josh followed Cephas up the winding stairs to the balcony of the beach house. "You know, Uncle Cephas, I still don't understand why Jonathan wants to sell this sweet beach house."

"Because he must move to Shreveport in order to help with my guardianship of you." Cephas opened the sliding door and led him across the huge den. He picked up a dishtowel and wiped the sunscreen off my nose. "And, we can't maintain two houses like this one and Ketrick's old house."

After Judge Bolton ordered Jonathan Steel to take Josh to Shreveport to live, Cephas had discovered he had purchased the house of the man who had been in league with the thirteenth demon. Such delicious irony? Or, God's strange sense of humor? It was in that house they had found the Ark of Chaos that led to their encounter with the eleventh demon and the evil assassins, the Vitreomancers.

"I'd rather he sell Ketrick's house. Dude, I can't remember much about the time when the thirteenth demon possessed me but that house is way too creepy. Except for the awesome pool out back." Josh snapped his fingers. "Uncle C, with my birthday coming up maybe I could have a pool party."

"Who would you invite? We've only been there two weeks." They had only lived in Ketrick's old house for a short time before the eleventh demon had surfaced along with the Vitreomancers. And, of course, Vivian. Darbonne. Ketrick. Wulf? Keeping track of her current last name proved to be as difficult as getting Josh to help with the packing. Cephas couldn't keep all of the woman's last names straight. She started out as Ketrick's assistant and orchestrated his defeat so she could move up the corporate ladder of evil. Then, she helped Rudolph Wulf with his vampire clans and ended up inheriting his empire. Just a couple of weeks before, she had shown up at Ketrick's house looking for her "Ark of the Demon Rose" as she called the antique box hidden in Ketrick's basement.

"I will allow you to have a pool party for your birthday. Perhaps you can invite Dr. Washington and . . ."

"Uncle C! You're right! She has some hot babes that work with her at the old church dig."

"College age girls. You have yet to enter the eleventh grade. Now, pack! Hurricane Leo put a scare in Jonathan. He was lucky there was no damage. So, while the market is hot, it is time to sell." Cephas wrapped a figurine in newspaper. Josh pointed to a pile of books.

"Why does Jonathan want all these books? They belonged to those other people."

"Those 'other people' were the only family Jonathan can recall before he met you, young man. The Pierce's took him in and gave him a second chance."

Steel had worked for Dr. April Pierce and her father, Robert Pierce after he had awakened on the beach two summers previous. He had lived in this house until tragedy struck. Josh glanced up the inner stairway toward the second floor. "Hey, Uncle Cephas, have you been in the room?"

Cephas paused and a chill ran down his spine. "I did not know him when he lived here in this house. But, it was in that bedroom that the killer possessed by the thirteenth demon did his dirty work. Jonathan said it is off limits."

"Is that where they died?"

"Tape that up." Cephas placed the figurine in a box and pushed it toward Josh. "Jonathan said Richard Pierce was attacked in the second floor bedroom."

"Yeah, nailed to the wall." Josh stretched tape across the box.

"But, April died on the third floor balcony. That is why he has lived down here on the first floor since April's death." Cephas placed more books in another box.

"So, what is he going to tell whoever wants to buy this house? Dude, is he going to show them what's in that room?"

"There's nothing in there but painful memories." Cephas

paused and glanced at the young man. Josh stared up toward the bedroom. His hand drifted to the scars on his chest and Cephas knew that deeper down, his heart ached with the loss of his mother. "Josh, I know you have suffered a great tragedy in the loss of your mother. But, Jonathan has lost three people he loved. I think it is best to leave the door closed on that part of his past."

"I don't think you understand, Uncle Cephas." He whispered.

"I do understand, Josh. Your mother was my niece and it is my fault she came to Lakeside."

Josh's iPhone rang. "Hello? Oh, hey Theo. Did you find the old woman's husband?" He paused. "You found her Poopems? What's a Poopems? A dog? What? Dude, what happened?" Josh's face filled with concern and he ran a hand through his hair. "He fell through the ceiling? Is he okay? Tell him I'm ticked off." Josh's face reddened. "Well, when he wakes up, tell him!" He shouted and tossed his iPhone on the coffee table. "Uncle C, I can't take much more of this. Jonathan got hurt again. And I bet there's some nasty demon waiting around the corner for us to fight."

"With what Jonathan has been called to do, you can expect many such times as this." Cephas said.

"Called?" Josh shook his head. "Jonathan hasn't said anything about a calling."

"He just hasn't admitted it yet. Helping others battle the evil in their lives is his destiny. It is his purpose."

"What is my purpose? Get kidnapped by another demon? Bro, what horrible death is waiting for me, now?"

Cephas put down the tape and crossed the room. "Josh, sit down." He pressed him down onto the couch. "The world is awash in evil. It is all around us. Unfortunately, you are close to the epicenter of a struggle between these two forces. It is true that being in close proximity to Jonathan Steel exposes one to the eventual battle with evil. But, I am convinced that you are also far safer when you are close to Jonathan Steel. He promised your mother he would protect you from evil. And, he always keeps his promises. He'll protect you, and if need be, others who are caught in the crossfire." Cephas gestured toward the windows. "Even those warm girls down there."

"Hot girls, Uncle C. Get it right." A hint of a smile then crossed Josh's face.

"Son, in your case, they will only be lukewarm."

Josh smiled and the doorbell rang. He hopped up from the couch and hurried across the den and through the kitchen. "Jonathan?" He hurled open the back door. But, instead of Jonathan, a woman stood there. Her blonde hair was pulled back through a cap. She wore a long sleeve running shirt and shorts. "You must be Josh."

"Who are you?"

"Renee Miller." She pushed past him into the kitchen. "I have to see him right now!"

"You're the one who let Jonathan and Theo hitch a ride on an airplane to Transylvania. Jonathan is not here right now."

Renee Miller froze in the middle of the floor and pointed a shaky hand at Cephas. "You! I should have known you'd be here. I hope you're happy, old man! This is all your fault!"

Cephas held up empty hands. "What did I do?"

Chapter 5

"Demonology is not just another crackpot-ology. It is the ancient and scholarly study of monsters and demons who have seemingly co-existed with man throughout history."

John A. Keel, UFOs: Operation Trojan Horse

The hearse drove slowly down "Maimed Street" past the Witches' Brew Tea House and the Necromancer's Nook. Vivian Ketrick looked through her reflection in the hearse's window at the passing shops of The Devil's Playground Amusement Park. Thunder echoed down the deserted street and lightning flashed against the broken windows and empty shops. The hearse bounced through another pothole in the weedy main street of the park and she bit her lip. Her demons swirled within her at the taste of human blood.

"Bile, do you have to hit every pothole?" She said. She dabbed her lip with the back of her hand.

The man driving the hearse glanced at her in the rear-view mirror. Fire filled his eyes and rather than meet that gaze with defiance, she looked away. Bile was not the incompetent wimp she had hired after the fiasco with Wulf. His normal sleepy eyes would reveal the thing that lived within him and if there was a spiral tattoo around his right eye, it might lash out at her. Again.

Vivian's demons were silent. Cowards! At the end of the disaster with the Ark of Chaos she discovered Bile hosted her old partner in crime, the thirteenth demon. Vivian touched the bruise on her cheek where Bile had struck her. She dared to look back at the man's eyes reflected in the rearview window. The fire in the man's eyes faded and there was no dark spiral tattoo around his eye. At the moment, the thirteenth demon wasn't at home.

"Sorry, Mrs. Vivian. This park has been abandoned for twenty years. No upkeep."

Vivian exhaled and realized she had been holding her breath. She wiped beads of sweat from her upper lip. Remain calm, she told herself and her demons. Tonight would be the most important of her life.

"They're doing this to humiliate me. The Council of Darkness couldn't possibly call this place home." They passed shops advertising all manners of witchcraft, Satan worship, dark magic, vampire clothing, and potions. No buttery beer in these shops! The broken windows reflected back the occasional flashes of lightning. Wind gusted around them and stirred litter and grit as the storm moved across central Florida at the tail end of Hurricane Leo. They arrived at the huge castle at the center of the park. No fairy tale princesses lived in this castle. Four passages branched off from the hub in front of the castle. A sign indicated the four divisions of the park: Underworld, Limbo, Gehenna, and Tartarus.

The hearse pulled around the circle and stopped in front of the castle. Vivian rolled down her window and the moist wind stirred her shoulder length hair. She had chosen a maroon evening gown the color of clotted blood. Maybe she should have worn a safari outfit!

"Florida is full of amusement parks, Bile. But, who thought anyone would go to a park based on Hell?"

"We would." Bile glanced at her in the mirror and his watery eyes glittered with malice. The spiral tattoo had reappeared around his right eye.

"You're back."

"I come and go. Now, get out of the hearse." The thirteenth demon growled.

Vivian opened the door and wind whipped around her, tossing her hair into a mess. Dirt and trash stuck to her dress.

The broken, black stones of the castle loomed above her. Towers and minarets stood against the stormy sky with crooked and deformed shapes as if sculpted out of melted wax. Windows sat at odd angles. Buttresses jutted out with statues of humans impaled on their spires. Human skulls sat at the top of spires that lined a drawbridge leading into the castle interior. On the castle's central tower, a winged demon crouched with huge unfurled wings. The infrastructure of the wings was broken and they dangled at an odd angle. The face of the demon was missing the eyes and the lower jaw. Bile appeared beside her. His stringy, white hair writhed wormlike in the wind. The spiral tattoo, the mark of the thirteenth demon, pulsed around his right eye.

"I know you planned on using the contents of the Ark of Chaos to influence the Council. Don't worry. It is safely hidden and I will show you how to control the council. For now, Vivian, our goals are the same. I can always kill you later."

Her anger surged bolstered by the newest demon she had acquired from the dying deputy and his Vitreomancer demon. Bile didn't know about that particular demon. "We will deal with thirteen, wench!" The demon whispered in her mind. It's about time you showed up, she thought.

"Angry, love?" Bile asked.

"Yes. I am here to demand a seat on the Council. I'm not here to do your bidding!"

Bile ran his red tongue over his fangs. They had once been fake. Now, they were real. He leaned toward her and a sudden viselike grip of invisible forces squeezed her throat. "I used to hate the taste of blood. But, the thirteenth demon has taught me how to cherish it." She could see the fine red veins in the white of his eyes and smelled copper on his breath.

"Release me!" The Vitreomancer demon screeched in her mind. "I will eviscerate this scum!"

Vivian had seldom felt fear in her life. Her demons gave her a measure of power but being caught between these two creatures overwhelmed her. The thirteenth demon was the worst of the worst; the most powerful and uncontrollable demon on earth. But, the Vitreomancer demon had shown her a history of unspeakable horror that threatened to eclipse that of the thirteenth demon. No wonder her other demons cowed in silence.

Bile's eyes widened and he stepped back. "What's this? You have a new tenant? How sweet! I cherish the coming conflict." He grabbed her chin with a cold, moist hand. "But, pay close attention, Vivian. Control your little demons, or I will destroy you and send them packing to the depths of hell. Do you understand?"

"Back off!" Her words were slurred by Bile's grip. "I told it to back off."

Bile smiled and released her chin. He leaned close to her and kissed her cheek with dry lips. "Don't be nervous, Viv." He whispered against her skin with cold breath. "Just realize that we will continue to destabilize the Council, not make it any stronger. That is why I am allowing you to become a member of the Council. Here is your task for now. Make friends with number ten and stop his plans. Understand?"

"Yes!" Vivian tried to straighten her tossed hair and brushed the trash from her dress. She stared into the dark tunnel that led into the castle. Deep inside, burning torches had been placed in brackets on the wall. The flickering torch light illuminated creatures carved into the wall. "I will do as you say for now, Bile. But, know this. One day, I will own this Council. And, when I do, you will bow before me."

Bile laughed out loud, an insane, high pitched giggling sound that echoed down the empty Maimed Street behind him. Lightning flashed and thunder crashed. Rain fell in huge gobs of water and Vivian was instantly soaked to the skin. Bile slid back into the hearse and rolled down the window. "Hurry, now, Vivian. I don't want you to catch your death of cold."

She rushed across the drawbridge into the castle's tunnel and ignored the things slithering in the thick, dark water beneath the bridge. Writhing, tortured souls covered the walls of the tunnel; naked men and women intertwined in agony goaded by horned and clawed demons. Deeper into the entryway were more carvings of those hacked and sawed in pieces by other humans in the service of the demons.

At the end of the tunnel, black candles flickered from sconces and the tunnel opened onto a huge ballroom. Back in the day, it had been a restaurant. Now, the interior had been cleared. The old tables and chairs had been pushed haphazardly to the sides leaving the center of the room available for the huge circular table with an open center. Twelve chairs surrounded the table each with a canopy over the top and a sheer curtain obscuring the chair's occupant. The canopies on two chairs opened to empty interiors. Twelve and Eleven once sat in them. But that was before they met her. Before they met Jonathan Steel. Don't mention that name! The Vitreomancer screamed in her mind. He is an abomination! Vivian ignored the voice and tried to see through the canopies but the light from three flickering globes hovering above the table only painted the canopies with shadows. To the side of the chamber near the discarded restaurant tables, Vivian noticed a portly man with huge white sideburns and a gleaming scalp glaring at her through a pair of small glasses perched on his nose. He was attired in a long, flowing gray robe. He stood behind a low table at which sat a bald young woman. Her eyes glowed with an unearthly blue light. Before her on the table sat a huge egg shaped object opened like a book. Flickering lights played across her face as she gazed on the interior of the thing.

"Good, record it all, my sweet." The man whispered to the girl. "Let the Grimvox work through you."

"Well, looky what the cat dragged in. Please come to the center of the table." A hoarse voice issued from the chair nearest Vivian. Before her, a portion of the table disappeared allowing her to step into the center of the table. She drew a deep breath and raised her chin in defiance.

"I will do as you say for now. But, know that I have defeated thirteen, twelve, and eleven and I claim their territories."

A deep-throated chuckle came from her right. It built in volume and was joined by laughter that filled the dark air.

"Don't laugh at me. You have already embarrassed me

with the hearse and demanding that I meet you at a deserted theme park." She took her place in the center of the table.

"We could snuff you out with a thought, brazen wench." A woman's voice came from one of the chairs. "Now be quiet or we will feast on your entrails."

"You've got to be kidding?" Vivian said. "You sound like a bad horror movie. No wonder you come here to play in a child's nightmare."

"We chose this place," another voice came from her left, "because we place you on the same level as a child."

Vivian whirled and tried to look through the veil. "I did not come here to be chastised by this council. I came her to become its newest member. Do you accept me or not?"

"It is not that easy." Another male voice came from behind. "You have shown great resourcefulness, Vivian. But, cleverness aside, you do not have the history sufficient for a member of the Council of Darkness."

"Our plans have unfolded for centuries." Another female voice echoed. "The plans of the master are convoluted and involved. They go far beyond your puny mind's capacity to understand."

"And, yet, in a few short months I have managed to thwart the plans of three of your most powerful demons." Lightning flashed across the room and thunder rattled the broken windows. For a second she saw the form of a squatty man in one of the chairs. "And, stopped the plan of your rivals."

"I like you." A man's quiet voice came from her right. Something stirred behind the curtain. "There are many factions in competition with this council. But, all such endeavors will end in failure. There is only one Council of Darkness. You have shown great resourcefulness and I, for one, think the Council needs new blood."

"You would, old man." The hoarse voiced woman said. "We do not have room for impatience. Our demonic powers are difficult enough to tame without allowing such impulsiveness to enter our midst. The master's plans are long in execution. We cannot lose sight of that."

The man's voice continued. "And, where have these plans

gotten us? The twentieth century was our playground, but here in the twenty first century, the enemy is gaining ground in spite of our successes. And, the times grow short."

"Oh, you always play the apocalypse card." Another man shouted from behind Vivian. "Enough of this sparring. I am the Master of this Council and I decree that this woman and her dominant demon be given a chance."

"She will have to earn it." The hoarse voiced woman said.

"I have earned it." Vivian propped a hand on her hip. "And, if you stand in my way, I will send each of you to hell and take your place just as I have three of your demons."

"Good!" the Master of the Council said. "You show initiative. We have needed such for a long time."

"It is a mistake." The woman's voice hinted at a Scottish accent from behind another veil.

Vivian drew a deep breath and slowly surveyed each chair. She planted a hand on her hip and examined her fingernails. "So, this is the mighty Council of Darkness? You are nothing but a horde of petulant children arguing among yourselves. Truth is, you are missing members of this Council. Your numbers are decreasing." She paused and let the Vitreomancer demon augment her voice. "And, as you have acknowledged, more powerful foes who serve the master grow stronger while you play games." A profound silence fell.

The Grimvox master chuckled and whispered to his assistant. "Quibble, she speaks of the Vitreomancers. This is excellent, my precious. Continue to record. I wonder if she has met the --"

"Silence, Keeper of the Grimvox. You are here as an observer only." The master of the Council said. "Vivian, I will forgive your defiance but only for one reason. We have a task for you." The curtain around one of the chairs opened to reveal an empty seat. "One of our members did not bother to attend tonight's meeting. He has betrayed us. The tenth demon devises a plan of personal advancement over that of the master's plans. You must stop him from realizing his final plan."

She clenched her fists. "I have done enough of your dirty work."

"You think that your victories over these three demons were yours to claim. But, you had help." The woman with the Scottish accent said. "The one known as Jonathan Steel."

"Jonathan Steel! I hate him!" Vivian shouted.

"Oh, so emotional." The woman said. "So childish!"

"I used him." Vivian whirled. "He was my tool. The plans were mine!"

"You used a blunt instrument when a scalpel would have been better." The woman said.

"Jonathan Steel is not a scalpel." Vivian stopped and concentrated on the source of the woman's voice. "He is brutish and vile and violent. He is a shark in a pool of minnows and all I had to do was stir a little blood in the water."

Behind one of the veils, a flame flared as if from a cigar or a cigarette. For a brief moment, she saw the outlines of a face before the light faded. "Ah, yes, Jonathan Steel."

"You have no say in the matters of Jonathan Steel." The woman with the Scottish accent said.

"Nor do you in the matter of Raven." The man said. Smoke drifted above his chair. "We all have our favorites, my dear. This Council realizes we have need of Jonathan Steel. Vivian, you cannot use him this time. And, you cannot harm him. We have plans for him."

"*You* have plans for him!" The Scottish woman growled. "And, Raven is dead."

"Jonathan Steel! Again!" Vivian said. "I am so sick and tired of that name."

"Good! Let your anger grow!" The Vitreomancer demon goaded her. "You can turn this to our advantage. We can take this Council. With me as the leader of your pack of demons, we can rule over all! Remember Bile's plans?"

Yes, I remember the plans, you troll! She hissed at the demon.

"Witness the inner struggle. We can sense it, Vivian. You cannot even control your own demons." The Scottish woman said. "How can you ever take a place at this table?"

"I now control seven demons. Seven! What ordinary human can make such claims?"

"Mary Magdalene." The Keeper of the Grimvox said.

"And, we know what happened to her demons. He sent them --"

The curtain in front of the woman with the Scottish accent bloomed outward and a bolt of blue lightning shot across the table. It hit the Keeper in the forehead and knocked him backwards into the stacked chairs. Quibble never moved and continued to stare into the Grimvox. "Enough of your interruptions! Never speak *his* name in our presence."

"And who is now out of control?" Vivian laughed. "You're no different from me. Everyone of you was once me. Remember that!" She licked her lips and tossed her hair out of her face. "All right, sweeties, I accept your assignment. But, don't underestimate little old me. I will sit on this Council. And, one day, I will rule it." Murmurs broke out around the table.

"Silence, everyone!" The smoking man shouted. "She has fire and vigor and ambition. Most of us are tired and old. Give her and her demons a shot at ten. Let us see if she can keep the tenth demon from realizing the insane plan of his."

The Council's master spoke up from his chair. "Very well, Vivian. We tire of you. Go! It is time for us to feast and no one outside the Council may witness a feast." He clapped his hands and movement around the periphery of the room caught her eye. Curtains hiding doors pulled aside and servants clad in black dresses or tuxedoes appeared. They pushed long serving carts covered with silver cloth.

A servant passed near to her and turned his zombie like gaze in her direction. His face was pocked with sores that ran with pus. His eyes were totally black. The silver cloth moved from something writhing beneath. The sound of moaning filled the room from whatever lay beneath the cloths.

She turned imperiously and walked right through the substance of the table. It took every ounce of demonic power at her disposal but she managed to make it to the tunnel before stumbling. Bile waited by the hearse. The rain had abated and left behind the fetid odor of decaying vegetation.

"Well, that could have gone better." Vivian said.

Bile smiled and the spiral around his right eye pulsed. "Don't worry, Vivian. I will help you." He reached into his shirt pocket and took out something that gleamed in the meager moonlight. "The talisman of the tenth demon."

Vivian held out her hand and he placed a gold ring with a bright, red jewel in it. "This stone once was seated within a medallion. But, I've had the stone set in a ring. Number ten won't recognize it."

"What do I do with it?"

"Are you familiar with the Grimvox?"

"The egg like thing?"

"Yes. It is the repository of all of our history and knowledge. It is our bible. I have taken key elements of number ten's history from the Grimvox and placed them in the talisman. It will inform you and it will empower you. At the opportune time, it will tell you number ten's true name and when you call him by his name, you will have power over him. So, use the talisman wisely."

Vivian slid the ring loosely onto the fourth finger of her right hand. It shrunk and fit snugly to her flesh. "Is this a ruby?"

"No, Vivian. It is a shard of the Bloodstone."

"Bloodstone? What is that?"

"In time, my dear, in time. Now, I must go and attend to some matters regarding number nine."

"But, what is it I am supposed to stop ten from doing?"

"Just pay a visit to Area 613 and you'll meet number ten." He tossed her the keys to the hearse and disappeared into the night air. "You'll figure it out."

"What is Area 613?"

Chapter 6

Conspiracy theories propose that some group of people is working behind the scenes, secretly orchestrating situations for their own benefit and the harm of others.

Mark Clark in "Lights in the Sky and Little Green Men"

"Looks like we got company." Theo said as he pulled into the driveway of the beach house.

Steel sat up and tried to focus as the pain medicine wore off. "Who?"

Theo pointed to the long, black limousine. "Not sure, Chief, but I'd say they have style."

Steel climbed out of the car and winced at the sudden lance of pain from his rib. He walked past the limousine and tried to see inside. His reflection stared back, face drawn with faint tracks of dried blood from the cut in his scalp. He followed Theo up the stairs to the back entrance of his beach house. The sound of angry voices came from inside. Theo threw open the door. Cephas stood rigid and unmoving in the kitchen faced off against a woman.

"Renee?" Steel said.

Renee Miller stopped and looked at him. "Where have you been?"

"At the hospital." He closed the door on the stifling heat. "What are you doing here?"

"I need your help."

Cephas stiffened and turn away. Steel pointed at him. "What's up with you and Cephas?"

"Bad blood." Cephas said as he slid onto a bar stool at the kitchen bar.

"We have issues." Renee crossed her arms.

"Then I have issues. You were screaming at each other."

Josh walked out of the bathroom and looked back and forth between Steel and Cephas. "Dude, what did you do this time?"

"I fell."

"Through the ceiling from the attic." Theo said. "Rescued a little poopy dog, though."

Josh rubbed his sunburned chest. "Dude, I'd like to get through one day without a major 'incident'. Is that too much to ask?"

"Then go back to Shreveport." Steel barked. Stop it! There was already enough anger in the room. He held out a hand toward Josh. "Wait! I didn't mean that. I'm on pain meds."

"Bro, that should calm you down." Josh pushed past him and collapsed on the couch. He started playing a game on his phone. "Maybe you should give some of it to Uncle Cephas."

"Renee, why were you two going after each other?" Steel asked.

"We have a history." Renee said.

"Details?" He leaned against the kitchen counter and fought off light-headedness.

"Renee was my student when I taught at NYU." Cephas said. "That was over ten years ago. She was bright and energetic and enamored with a certain young pilot who had formed his own charter airline."

"Robert Miller?" Steel asked.

"Cephas tried to warn me." Renee crossed her arms. "He knew something about Robert and I wouldn't listen."

"And?" Steel glanced at Cephas.

"Jonathan, you know that I collect artifacts. Robert Miller had come to me the year before asking if I knew anything about the Tablets of Anak."

The hair on his neck rose. He glanced at Theo. "Anak, again."

"Don't look at me, Chief. I ain't the boss of coincidence." He said.

"What coincidences?" Cephas asked.

"Later. Anak?" Steel prompted.

"Yes." Cephas rubbed his huge mustache and shrugged. "A king from the Bible. His sons were the men who were seen by the spies that Moses sent into the Promised Land. You remember the story?"

"I haven't had much time to read the Bible lately and, if you remember, I have a problem with my memory."

"Of course. After leading the children of Israel out of Egypt and through the wilderness, Moses sent some spies into the Promised Land. The people who followed Moses wanted to take back the land that had been theirs before they ended up in captivity in Egypt. The spies came back with stories of powerful men and giant plants. Most of the spies were frightened and discouraged."

"Except for Joshua and Caleb." Josh said. Steel glanced at him and he shrugged. "What? Bro, that's why my mother named me Joshua. Go figure."

"The powerful men the spies saw were the sons of Anak." Cephas said.

The world closed around Steel, narrowing his life and focus down to another demonic encounter. First, a talking dog spouting information about the sons of Anak. Then, Major Miller and a stranger who somehow knew Josh. And, now, Renee was here at the condo. "So how did this figure in with this Major Miller?"

"He searched for a legendary urn containing vials of an elixir used by Anak and his sons to give them great power and stature. Once Anak learned of the Israelites' possible invasion, he hid the urn. Its whereabouts were written on a tablet." Cephas said. "Now, Jonathan, I had never heard of this particular legend. And, something about Robert troubled me."

"You never did like him." Renee said.

"I didn't divorce him, my dear. You did." Cephas regarded Renee with a look of disdain. "God has given me the spiritual gift of discernment. I sensed immediately that he was somehow in the grip of evil." Cephas ran a hand through his bushy hair and gestured with his knobby hands. "I did my own research into the Major. He was involved with some type of quasi-military project. And, it looks like he still is."

"You are right about that. I met him at the hospital." Steel said.

"He's here?" Renee stiffened.

"He was looking for a fugitive."

"No! He can't be here." Renee said.

"Why?" Steel asked.

"He runs a black ops program." She glanced at Cephas. "But, not so much for the government anymore. He also received funding from some wealthy, powerful donor." She paused to catch her breath. "The donor is a collector of ancient artifacts. He hired Robert to find the Tablet of Anak."

"I am more concerned Miller is here for the children." Cephas said.

Renee froze. "No!"

"What children, Renee?" Steel asked.

"Vega, my daughter. She's missing." Tears filled her eyes and she pressed her hand against her mouth. "Oh my! Cephas, it can't be!" Renee turned her tearful gaze on Steel. "You remember that day in my office? You saw a picture of Vega."

"Yes." Steel nodded. He recalled the baby in the photograph had seemed, well, unusual in appearance. "Have you told the police?"

Renee shook her head. "It's complicated. That's why I need you to help me."

"Uncomplicate it for me."

"Robert and I had separated before Vega was born. When she was born eight years ago, she was a beautiful baby girl. But, within a year we knew something was wrong. I took her to several specialists. They had no idea about her condition. The stress was more than I could take and Robert and

I couldn't work things out. So, I took it out on him. I didn't know about his connection with the black ops group then."

"If you had, you probably wouldn't have taken him to the cleaners." Cephas said.

"You think all of this is for revenge?" Renee asked. "No, there is more. He wants to get his hands on Vega before I find a cure. Jonathan, I used part of the divorce settlement to get my Ph. D. and formed Ingenetics Laboratory to try and find a cure."

"And, you took his airplanes." Cephas said. "And, that, my dear is another reason he is out for revenge. Jonathan, he convinced the military to investigate Vega's disease."

"They wanted to experiment on her." Renee said. "Can you imagine it, Jonathan? Your own father wanting to experiment on you?"

"I can imagine, Renee." Steel's father had experimented on him by performing some kind of surgery on his brain to control his memory. At the first brush of the memory of that flashback, the nausea came and he pushed it away before the implanted post hypnotic reaction could kill him. He swallowed and cleared his mind. "How did you keep Vega safe all of this time?"

"Someone defected from Robert's group and decided to help us. He helped me hide Vega. He has kept his identity a secret to protect the children." Renee wiped away the tears. "We found twelve other children with the same disease. Robert had gone after each one. So, we hid them from Robert. Our guardian hired mercenaries to protect them and moved them from location to location and our last one was a school." She paced around the kitchen. "When Hurricane Leo came through Pensacola, it took out the school defenses. All of my tracking data and video surveillance went dead. I don't know what's happening at the school. Either they're in hiding or, worse, they're out there wandering around in the swamp."

"What about this person you hired to protect them?" Theo asked.

"I haven't heard from the Guardian. That's what we call him. Now you know why I can't involve the police. They'll lead Miller right to Vega. I called your house looking for you and Dr. Washington answered. She said you were here at your beach house and she was checking on your house. That is why I came to you."

"Is the school near Bagdad?" Steel asked.

"Yes. How did you know?"

"When Major Miller was at the hospital nosing around he was looking for someone from around Bagdad. Hopefully, it was your Guardian unless someone else is looking for the children."

Renee shook her head. "Only Miller's boss but he would be working through Miller. That man in the hospital could have been the Guardian. And, if he was hurt while looking for the children then Vega is unprotected." Renee's face blanched and she put a hand to her mouth. "Jonathan, if Robert gets these children, he will take them to some secluded lab and he will dissect them like lab rats!"

"How can an eight-year-old child survive in the woods?"

"Vega is, uh, resourceful." Cephas said.

"Resourceful? Define resourceful."

"She is special." Cephas said.

"Stop the riddles already!" Steel drew a deep breath and fought the growing fury within. "I just fell through a ceiling, cracked a rib, and was lectured to by a demon possessed Pomeranian. I'd like some straight answers!"

"She is advanced for her age." Cephas said. "She has enormous strength and agility for her age. A very high I.Q. And, special abilities."

"I thought Renee said she was sick." I said.

"We've tried to keep all of this secret, Jonathan. Don't get mad." Renee glanced at Cephas. He nodded for her to continue. "Vega's abilities come with a price. If we don't find a way to stop her development soon, she will be dead within a couple of years. That is why we must find her now, Jonathan." Renee's phone warbled and she pulled it out of her pocket. She frowned. "I need to take this." She walked out onto the deck, closing the sliding door behind her.

"Cephas, sounds like this Major Miller is dangerous." Steel watched Renee through the beach house windows.

"He has his own personal agenda, Jonathan. And, he has the military might behind him. If he finds these children, they'll be locked away forever."

Josh joined them at the bar. "So, like are these children powerful? You know, telekinesis? Mutants?"

"That I do not know." Cephas patted Josh's arm. "But, they are gifted enough to interest the government."

Renee spun around. Her eyes were wide with shock. She burst into the room. "We found it! We found it, Cephas."

"Found what?"

"The Infinity Disc!"

"Are you sure?"

"Fifty feet down in a hidden chamber. Right where I thought it would be!"

"What is the Infinity Disc?" Steel asked.

"It may be the answer to my prayers." Renee said.

"Now, Renee, you shouldn't put your hopes on another artifact. Remember the last one?" Cephas said.

"I know. But, if the rumors are true, this could cure Vega. And, all the children."

"Renee, I've heard a dozen different rumors about this thing. You need to bring it back to your lab and let me take a look at it --" Cephas said.

"You're not going anywhere near it." Renee backed away. "The last time you tried to help me with an artifact, the thing crumbled to dust." She paced across the kitchen. "I can get a transport and be there in a few hours. But, this has to be kept a secret. No one must know, Cephas. Especially my sister."

"Your sister?" Josh asked. "You have a sister?"

"Yeah, she and I aren't exactly on speaking terms."

"I met her at your office." Steel said. "She's a television star."

"The Artifact Hunter." Cephas rolled his eyes. "Making a living off of sensationalizing archeology."

"Dude, I've seen that show." Josh smiled. "She's hot! And, now that I think about it, she looks just like Renee."

"We are twins." Renee sighed. "As much as I hate to admit it. And, you can bet she'll want the disc for her television show. That's why I've kept this dig a secret. No one must know. I have to go. I have to be there to see the thing properly crated and brought to my lab." Renee slid the phone into the pocket of her shorts. She stepped closer to Steel and reached out and took his hands in hers. "Jonathan, I helped you with the twelfth demon. You remember?"

"Yes. And, I promised you I would help you in anyway."

"Now is the time. I need you to find Vega before Robert Miller finds her. Will you find my daughter?"

"A promise is a promise." Steel said.

PART 2

ANOMALY

Anomaly (AN 1-5): The observance of a UFO and other accompanying or similar abnormal phenomena (for example, poltergeists).

Flyby (FB 1-5): The observance of a UFO flying in the sky.

Maneuver (MA 1-5): The observance of a UFO that exhibits a discontinuous trajectory in the sky.

Jacques Valle's UFO classification system

Chapter 7

Some people claim that physical objects carry with them a hidden power or ability. Claims are made that such talismans can be used to augment human abilities. I've been investigating these kinds of artifacts for years and I've never substantiated that any object contained hidden power. If there is power attached to these artifacts, it comes from some other supernatural source. And, of course, I don't believe in the bogey man!

Dr. Cassandra Sebastian, "The Artifact Hunter" Season 2 Episode 5 "The Search for the Ark of the Covenant"

The Guardian begged a ride from a man with a trailer full of hay and rode from Pensacola to Bagdad to retrieve his motorcycle hidden beneath some palm fronds. He headed off down the highway back toward Pensacola. The afternoon was hot and humid and he had a couple of hours before dusk. He had much to do. He arrived at an outdoor storage facility and rode to the back of the complex. The compartment he rented was hidden around a corner out of view of the highway and the main office. He unlocked the combination padlock and slid the door high enough off the ground to bring in the motorcycle.

A small air conditioning unit in an air vent in the far upper right corner made the interior habitable against the sweltering August heat. He closed the door behind him and turned on the main light. Half of the compartment contained boxes of supplies. The other half had a makeshift camper's shower and an inflatable bed. A fold out table housed his laptop and a small printer.

"Okay, Miller is here and he doesn't know about the school." He paced around the small room. "But, he will find it. I need to make sure the children aren't hiding in the basement." He paused in front of a small mirror duct taped to the wall. "And, now, Jonathan Steel is involved. Jonathan Steel didn't recognize me and I need to keep it that way. It's time for a shave and a haircut."

He dug through one of the storage boxes and pulled out a hair clipper and shaved his hair down to a buzz cut and trimmed most of his beard. He stripped off the muddy sweats and stepped into the shower and opened the bag of cold water hanging from the ceiling. He had tapped into the irrigation sprinklers but hadn't been able to find a way to heat the water.

The cold water hit him in the face and he gasped. The cold water threw his sore muscles into spasm. He stretched and pulled the pain out of his muscles. A huge bruise covered his upper back where he had hit the water full force.

He showered and shaved away his beard. When he was finished, he pulled on a faded pair of boxer shorts and an old tee shirt and sat down at the laptop. His reflection in the screen showed a different man from the one who had tried to rescue Vega.

"Okay, so we had an encounter with a flying saucer. What's new? And, what was that thing with four wings? Where did it come from? If Miller has that kind of advanced tech, then Vega is in more trouble than ever before." He said. He hacked into a government satellite feed and sifted through the images taken over the Bagdad area the previous night.

"There!" he stabbed a finger at the image on the laptop. Against the cold, flowing waters of the Blackwater River, a small hot blob appeared on the infrared image. It drifted across the screen then disappeared in a flash. "So you were real and not just a dream. Even that fly-like creature was real." He tried to recall the details of the thing with four wings but his mind slid down a slippery slope to the edge of insanity. Thinking about the thing became difficult. Nausea gripped him and he smelled the sweet odor of the gas. He grabbed his head.

"What's happening? What kind of drug did you use on us, Miller?" He pushed away the memory of the thing and his nausea subsided. He glanced back at the screen of his laptop. "But, if you were behind the abduction, why were you at the hospital? If you had Vega, then you were done. Why risk exposing your presence unless --"

He jumped up from the table. Was someone else looking for Vega? No! It can't be! Could *he* be here? His heart raced. "You want them, don't you? If he is here, then everyone is in deep trouble. Vega. Josh. Renee." The man shut off the laptop with a trembling hand and hurried over to the equipment container and began to construct a strategy as he pulled on dark sweats and stuffed equipment into a backpack. "You should have kept Josh in Shreveport, Steel. Now, I have two people to protect. Two!"

He paused and ran a hand over his sweaty brow. "Supernatural forces are bringing us all together. Why now? And, if he is involved I will need more than just guns and tech." He closed his eyes and shook his head. "No! I promised I would never use it again. I can't!" He glanced up at the ceiling. "God, I can't!"

His ragged breathing filled the chamber and against it he heard a faint pulsation coming from deep within the equipment case.

"No! Please?" He pleaded to the empty air. "I can't do this again. I can't!"

The thing in the case brushed his mind with its promises. From the deepest corner of the case, it called to him; dark whispers; red tinged promises of power; calling, calling. He paused at the end of case and stared at its carefully stacked contents.

"I have to save Vega." He whispered. "I have to save the other children. I have to save Josh. I don't matter. I gave up that option long ago." He surrendered to the allure of the thing. He tossed aside equipment bags and clothing until he found the pouch tucked innocently in the corner. It whispered to him; called to him; an unholy voice of great power and seduction. He took the pouch and placed it on the table next to my laptop.

"Forgive me." He groaned. He studied the pouch, a heart of darkness, death and destruction dwelling within. If he took it out; if he went back to that time would he lose himself? He drew a deep breath and unzipped the black leather bag and poured the necklace into his palm. A red jewel the size of a pecan pulsed briefly with energy and the shock of its touch sent electricity through his arm. The whispering stopped and it lay cold and hard against his skin. He hated it and yet loved it. He wanted to throw it across the room and yet he wanted to swallow it whole; to let it become his very being.

"I have no choice." He said to the empty room. He dropped the silver chain around his neck and let the jewel settle over his heart. Its evil touched his skin with a brush of heat and the echo of pain and death passed over his memory. He smelled mountain ice and ancient death dust. He tasted dried blood and pale, white sweat.

"God, be with me." He whispered as he slipped into the night.

Jonathan Steel stepped out of the trees and stood beneath a star filled sky. He reeked of mosquito repellent and sweat. For an hour, he had hiked his way down the trail Renee had described deep into the forest until he arrived at the isolated school. Before him sat a two-story antebellum plantation home with a wraparound porch at each level. Playground equipment littered the front lawn, tossed aside by the winds of hurricane Leo. A rutted road covered with gleaming white shells led up to the front of the building.

His mind reeled with the developments of this day. Renee Miller and her deformed child. The Children of Anak and a talking dog. It was happening again. Demons haunted his every thought and they were being pulled back into the battle. He desperately wanted to just walk away, to disappear into the woods and forget this day. But, he could not. He would not.

Steel placed his night vision binoculars to his eyes and studied the building. The only light came from two gas lamps on either side of the wide stone stairway that led up to the first level porch. To blend in with the night, he wore dark pants and a long sleeve black shirt with a ski mask pulled down over his face. Sweat ran in rivulets down his neck and along his back. He slapped at another mosquito and sprinted down the shell driveway to the foot of the stone stairway and up to the porch. In places, the plywood tacked over the windows to protect the school from the hurricane had fallen loose and broken glass glittered in the starlight. He glanced through the closest broken window into complete darkness. He struggled through the large empty pane and stepped into the dark, dank interior of the school.

Steel fished a pair of night goggles out of his backpack. The room exploded into view in shades of fluorescent green. Chairs and music stands surrounded him in what must have been the music room. The stands were all tossed on the floor and sheet music covered the chairs like snow. Wind-blown dirt covered a grand piano. He eased into the long hallway Renee said would lead down the center of the building.

Glass crunched far back in the rear of the building. Steel froze, his senses alert. Someone was here! Was it one of the children? At the other end of the hallway a figure moved through the shadows.

He calmed his breathing and stepped quietly down the hallway passing two more classrooms and a play area filled with easels. The smell of rotten food hung on the still air and he stopped just short of the door leading into the dining hall. He peeked around the edge of the door and as his goggles slid down his sweaty nose.

Pain exploded in the back of his neck and he fell forward and instinctively rolled to his right. Something metallic banged against the tile floor where he would have landed. He spun on his backside and with his right leg, raked the other person's feet from beneath him. The figure bounced upright immediately and disappeared down the aisle between two long dining tables.

Steel's dizziness cleared and he hurried through an open door into the darkened kitchen. His feet caught the edge of a serving cart and metal utensils rained down upon him. Knives, forks, and spoons splattered on the floor and the other man slid on them and fell forward onto the serving cart. It rolled across the kitchen and into a counter. Pots and pans cascaded down onto the attacker and filled the silence with a raucous cacophony. Steel shrugged off the utensils and pulled himself to his feet. The other figure threw aside the pots and hurled a roasting pan. Steel ducked and tossed a handful of forks at the man. He heard him gasp in pain.

Steel dove beneath a metal preparation table even as the other guy vaulted over the top. The man's feet slid on the utensils and he banged up against the metal door of the walk in refrigerator. As he fell down, he grabbed the handle and the door burst open on a warm wave of fetid air. The smell of rotting meet and spoiled dairy products choked Steel. He pushed the serving table and it tumbled over, carrying with it a rack of pans and large utensils. The other man screamed as the table fell on top of him.

Steel climbed out of the mess and adjusted his goggles. The man's legs stuck out from under the preparation table. He grabbed the man by the feet and dragged him out of the kitchen through a rear door and out onto the back porch. Steel blew the odor of rotting meat out of his lungs and sucked in fresh air. The man groaned and started to move. Steel kneeled on his chest and pushed down the man's arms. His face was covered with black paint and a similar pair of night goggles was askew on his eyes. Steel pulled the goggles away.

The man had cut his hair and shaved off his beard and now Steel recognized him. He had seen the man's face in a photo on the refrigerator of Josh Knight's house in Texas. The man was Josh's father!

"You're supposed to be dead." Steel said. Arthur Knight blinked and his eyes focused. He tensed as he tried to fight back. "Don't even think about it. You've got a lot of explaining to do."

"Steel? I thought you were with Miller." He relaxed. Steel leaned back and Knight flipped him backwards, spinning to land on Steel's chest with his knees. "I don't have time for you."

"Does Josh know you're alive?" Steel said hoarsely.

Knight stiffened at the mention of the name and Steel took the opportunity to roll from under him, pinning Knight against the floor once again. "We can do this all night or you can stop fighting me and tell me what's going on. Truce?"

Knight nodded. Steel released him and stood up quickly. "I guess you're the Guardian?" He helped the man to his feet.

"Yeah, some Guardian I turned out to be. I'm here to find the children."

"Vega?"

Knight glared at him. "How do you know about Vega?"

"From Renee Miller. She sent me to find Vega."

A reddish bright light burst from somewhere over the top of the school. A flare floated high in the sky.

"They've found the school." Knight pulled Steel back into the shadows of the porch. He pointed into the kitchen. "You can bet they've got the building surrounded but I know a way out of here."

"I don't trust you." Steel said.

"The feeling is mutual." Knight's eyes glowed against his dark face paint. "But, we both have Josh's best interest at heart. Remember that. Now, follow me." Knight hurried into of the kitchen and down a back hallway.

"Who is out there?"

"Miller and his men. If you want to get out of this alive, you do exactly as I say. Understand?" Knight pointed toward the far side of the main hallway. "This way." They went through a classroom to a door that opened under the stairs in the center of the building. Already, lights were bouncing around outside with the approaching group of men.

"If the children are still here, they'll be in the basement." Knight opened the door to another set of stairs that led down into darkness. Steel pulled the door shut behind them. He followed Knight down into the stuffy, musty air of a basement. The ground was covered with dirt and brick and Knight headed down a corridor of broken brick and mortar. They heard the crash above them as the men burst through the front door, then muffled shouts as orders were given and the sound of scurrying feet.

"Look." Knight pointed to the dirt. "This is how they left."

The shallow dirt of the basement covered a rough brick floor and in the dirt were the imprint of huge bare feet leading to a blank wall. Knight tapped on the bricks and pushed and prodded. "They went through the hidden passageway."

Steel helped him probe until a section of the bricks moved and a narrow sliver of the wall swiveled away allowing the odor of swamp water to fill the hallway. Steel followed Knight and pressed the door shut behind them. They stood on a wooden dock in an underground chamber. Water led away down a tunnel.

"They swam out of here."

"What is this?" Steel asked.

"An underground escape used during the Civil War to get slaves out of this area to the river. We've got to hurry. They'll find us soon." Knight jumped into the water and began to swim.

Steel looked once toward the brick façade behind me. Hanging on the wall were rusted metal boat hooks used to push and pull longboats in the water. He took one and wedged it across the door. He eased into the murky water and something glittered near his feet. He reached down and pulled the silver chain from the water. A jewel shaped like a drop of blood hung from the chain. For a second he thought it pulsed with a weak light. As he gazed into the depths of that jewel, the world seemed to fade around him. A shadow of evil eclipsed his vision and he almost dropped the thing back into the water.

"What?" His usual cloud of paranoia and anxiety seemed to lift for a second and for the first time in memory he felt normal. He drew a deep breath and emotions gripped him, an undulating mixture of deep fear and complete contentment. What was this thing? His reverie was broken by the sound of approaching soldiers. He slid the jewel into his pocket and swam down the tunnel.

Starlight loomed ahead of him at the end of the long tunnel and the waterway joined a stream. He slid out of the tunnel and glanced around. Arthur Knight was nowhere to be seen. In the far distance, he heard voices and the unmistakable sound of a helicopter. He climbed up the slope of the creek and into the trees. Bright lights illuminated the lawn and soldiers bustled around like ants disturbed from their work. A helicopter touched down and Major Miller stepped out into the red light of the flare. Things were getting complicated. Just like always. He set off after Arthur Knight.

Chapter 8

"I should have worn sunscreen." Vivian shaded her face against the harsh desert sun. She studied the uncaring gaze of the security guard posted at the entry gate for "Area 613". "I called and talked to someone. I am expected. Vivian Darbonne Ketrick Wulf. Check your logs again."

The guard shook his head. "You're not on the list."

Vivian swore and leaned against her rental car. Where was Bile when you needed him? She winced and jerked her hand away from the hot car surface.

"Let me take him." The Vitreomancer demon spoke in her head. Since acquiring the demon, her other demons had fallen silent becoming subservient to this new demon. Strangely, she missed the shark demon and Summer, the seductive vixen. She would be helpful right now.

"No! Let me handle this, Vitreo." Vivian hissed in her mind. She looked at her hand and winced. She blew on her palm and held it out toward the security guard. She gave him her best painful look. "Officer, sweetie, I think I burned myself on the car hood. Do you have any lotion?"

The security guard remained silent and unmoving when his gaze shifted from Vivian to the road behind her car. A white limousine halted in a whirlwind of dust. The cloud continued and enveloped Vivian. She coughed and tried to keep the sand off of her dark dress.

"You need to pull out of the way so I can let this car in." The security guard said, unmoved by the dust cloud.

"What makes that car so special?" Vivian said hoarsely. She coughed and gazed through the haze. A door opened and a woman stepped out of the vehicle. She was tall and wore a three-piece white suit over a maroon silk blouse. Her blonde hair was cut short. Turquoise earrings dangled from her ears.

"Is there a problem, Brandon?" She said as she walked toward them. She paused and pointed a perfectly manicured finger at Vivian. "Do I know you?"

Brandon smiled. "Sorry, Dr. Sebastian. I'm just trying to get rid of an unwanted visitor."

"Yes, we have met." Vivian glared at the woman. "The Artifact Hunter. The last time I saw you, your audience was covered in blood."

Dr. Sebastian raised an eyebrow and chuckled. "Yes, I remember you now. Victoria? Valerie?"

"Vivian."

"That's right. Robert Ketrick's secretary." She nodded. "Didn't you marry that creep right before he died?"

"I did." Sweat trickled down Vivian's back. "And, Rudolph Wulf. Area 613 houses artifacts?"

"No, Vivian, I am an investor in Area 613. Now, if you would be so kind as to move your rental out of my way, I would like to get on with my business."

"With?"

"None of your business, Viv." Dr. Sebastian said. "You are a long way off from your precious vampire clan, aren't you?"

"I'm here on business also."

"What kind of business?" Sebastian asked.

"Remember out last encounter? I'm here on that kind of business."

"Supernatural?" Sebastian laughed. "Get out of my way. Now!" She gestured to the security guard and he reached for her arm. Vivian pushed it away.

"Wait. You want to know the truth? I was sent here by the Council of Darkness to take down the tenth demon." She waited for a reaction. For all she knew, this woman was the tenth demon.

"The tenth - demon, did you say? Now, where have I heard something like that?" Sebastian drew a deep breath.

She glanced up at the sun. "Brandon, give us a moment. Vivian, in my limo while we sort this out. You have definitely intrigued me."

Vivian followed her back to the limo and gratefully slid into the cool interior. Sebastian sat opposite her and sipped at a bottle of water. "This demon you're looking for, does this have anything to do with Ketrick's appearance on my show?"

Vivian's finger warmed and she glanced down at the ring. The red jewel pulsed slowly and her memory of that night returned.

Quantum cerebral contact:
Grimvox Archive Vol:
3102:1123:17.45.16
Sebastian Studios: Sound Stage 13
Cue Excerpt 7

Vivian hated the woman the minute she laid eyes on her. Dr. Cassandra Sebastian was tall and athletic with long, blonde hair styled around her electrifying smile. She managed to make the khaki suit she wore look like a designer outfit. Vivian had expected a man on whom she could use her looks and body language. But, Dr. Cassandra Sebastian barely acknowledged her presence and dismissed her instantly to focus on Robert Ketrick.

"Bob, I have been waiting a long time to have you on my show." Sebastian shook her boss's hand. A loose turtleneck and dark blazer hung on his lanky figure. His long, black hair was braided into a snake like pony tail that hung down his back.

"I am a busy man, Dr. Sebastian. But, I am here thanks to my new administrative assistant, Vivian Darbonne."

"Without the apostrophe." Vivian said with her heavy Southern charm. She held out her hand. But, Sebastian's smile wavered as her eyes raked over her with the kind of look that told her she thought of her as nothing more than a tool; a thing used by Ketrick. Her admiration was for her boss only. Vivian dropped her hand and dreamed of tearing the woman's throat out and drinking her blood. Ketrick had just lured her away from her vampire clan and she still craved the coppery taste of blood. Her bat demon wiggled in anticipation.

"No, my love," the demon whispered, "She may be an arrogant fool but we might still have use for her."

"This way to the green room. You need some makeup. You look positively skeletal." Sebastian turned Ketrick away from Vivian and headed down the hallway. Another man stood in the shadows of the hallway and his gaze came to rest on her. He was as tall as Ketrick but more muscled and toned than her skeletal boss. He was bald and something about his face disturbed her. Ah! He had no eyebrows. His eyes glowed a shocking shade of lavender that stood out against his dark, milk chocolate skin. He stared at her without emotion on his seamless features.

"The organ grinder complained to his monkey. He chastised his pet for his poor dancing." The man said with a touch of a British accent.

"Pardon?" Vivian said.

The man stepped out of the shadows. "To which the monkey replied, 'If you would learn how to play the grind organ, then perhaps my dancing would be better.'"

Vivian lifted an eyebrow and smiled. "I get it. Who is in control?"

"Dr. Sebastian is a pretentious scam artist, you know." The man said in his strangely hypnotic accent. "Her love is for the limelight. But, I think your associate assumes he is in control of tonight's show. This will be an interesting diversion for me. Although, I am here to discuss an artifact Dr. Sebastian is seeking."

"I thought 'The Artifact Hunter' doesn't give up her artifacts. She takes them instead." Vivian nodded.

"I am Anthony Cobalt, Ms. Darbonne, was it?"

"Yes. Without the apostrophe."

"What happened to your Southern accent?"

"Well, we are perceptive, aren't we?" Vivian said. "Let's

just say I save it for the organ grinder. Aren't you the man who built the private space ship? The one that just orbited the earth with the first space tourists?"

"Yes." Cobalt blinked and Vivian studied his hairless face with fascination.

"I suppose you are intrigued with my appearance? I spent a long summer in a traveling carnival in Europe. I was fifteen at the time and had barely escaped the apartheid of my native South Africa. They called me the 'Hairless Wonder'. I met a young girl who took my fancy. She was called 'The Bearded Girl'. We complemented each other. Unfortunately, she jilted me for the 'Snake boy'. She liked his scales."

"The 'Hairless Wonder'?" Vivian shook her head. "Why are you telling me this?"

"Just prattling out loud. Vivian, you see, I suffer from congenital alopecia. I am what is known as a chimera. When I was in the womb, I absorbed my fraternal twin brother. The results were catastrophic for him but, shall I say, fascinating for me. I have two genotypes. They continual war with each other and somehow, my hair lost. Now, if you'll excuse me, I must go to the green room and prattle on with your boss. One takes publicity where one can get it, Vivian."

He walked away and the demon within writhed in confusion. "That man is one of us!" It said.

"Really?" Vivian replied. "Then we shall keep an eye on him. Perhaps you would like some company."

"No, my love. His demon is very strong. I would suggest we avoid him for now."

The camera light blinked and Cassandra Sebastian, aka, The Artifact Hunter smiled as her show went live. "Welcome back to today's episode of 'The Artifact Hunter'. I am Dr. Cassandra Sebastian, the Artifact Hunter."

The sound of a cracking whip punctuated her words. "In our first segment, we spoke to Anthony Cobalt and let's welcome him back." The audience applauded and Sebastian tucked her long, blonde hair behind her ears and turned in her seat to face Anthony Cobalt.

"I believe the women in our audience are curious, Mr. Cobalt. You have striking eyes. What color are they?"

"Lavender." Cobalt smiled. "A genetic curiosity."

"Indeed. Now, tell us again about this pin you wear." She gestured to a golden pin on the lapel of Cobalt's blazer.

"Orion was the son of the sea-god Poseidon and could walk on the waves. He was later blinded by Oenopion and journeyed to the uttermost East where Helios, the Sun, healed him. He soon went to Crete where he hunted with the goddess Artemis. In the course of their hunt, he threatened to kill every beast on Earth. Mother Earth sent a giant scorpion to kill Orion and after his death, the goddess asked Zeus to place Orion among the constellations as well as the Scorpion. My pin carries the belt of Orion, Dr. Sebastian. Orion was a hunter in mythology and I chose this emblem for my latest space vehicle."

"Really? I hope you don't plan on using your spacecraft to kill every beast on Earth." Sebastian said and smiled.

For a moment Cobalt looked grim and his gaze shifted to Vivian. Something about the look on his face filled her with dread. Cobalt smiled. "On the contrary. I merely wish to be the first private contractor to carry tourists into space."

"And, that is exactly what you accomplished last week." Sebastian nodded toward the audience. "Your craft, the Orion, managed to orbit the Earth last week how many times?"

"Over a dozen."

"With the world's first space tourists, ten fortunate individuals. I'm just curious, why didn't you join your guests on the inaugural flight?"

"Space sickness." Cobalt frowned. "Truly ironic, wouldn't you say?"

"Yes."

"And, I've heard from a little birdie somewhere that you are interested in ancient artifacts as I am. But, in space?" Sebastian said.

"There is a legend, Dr. Sebastian. I spoke earlier of

Artemis and it is said that Alexander the Great journeyed to the temple of Artemis to view the Jupiter stone. It was said to be an ancient artifact that once orbited the Earth until the gods threw it down from the heavens. I have not chosen to look for artifacts in the celestial sphere. But, I do long to find out what is out there beyond our world. That is why I have built my space station."

"The privately funded 'Eagle's Nest', right?" Sebastian said.

"Yes. There is a legend as a basis for the name of my space station. Depicted as an eagle, Aquila is named for the bird that belonged to Zeus. Aquila's most famous task was carrying the mortal Ganymede to the heavens to serve as Zeus' cup bearer. In the constellation Aquila, the stars Vega, Deneb, and Altair make up the summer triangle. Thus, the Eagle's nest."

"Dr. Cobalt, if you do find that Jupiter stone in orbit, please give me first shot at the exclusive unveiling." Sebastian said.

"I would consider no one else, Dr. Sebastian."

"Well, thank you so much for coming on the show." She turned to the television cameras. "I'd like to introduce you to our next guest. Mr. Robert Ketrick is a well-known businessman and I am quite envious of his fascinating collection of artifacts. Welcome Robert Ketrick."

"How do I look?" Ketrick asked as he prepared to walk out onto the stage.

Vivian straightened his blazer. "Fine, Robert. I picked this out myself."

Ketrick glanced at the huge cabinet beside Vivian draped in a red cloth. "My dear Vivian, this will be a night long remembered." He sauntered onto the stage and shook Sebastian's hand and sat in the seat next to Cobalt. "Good to see you again, Tony."

"Anthony, Bob. Please call me Anthony." Cobalt looked away and grimaced.

Ketrick flicked his long braid over a shoulder. "Mr. Cobalt and I are not on speaking terms, Dr. Sebastian. Territorial dispute, you might say."

"Really. I thought maybe it was the battle of the hair versus the, well, un-haired." Sebastian laughed.

"Yes, my hair braid is reminiscent of a scorpion." Ketrick glanced at Cobalt. "You know, like the one that defeated Orion."

An awkward silence descended while the two men glared at each other.

"Well, we could go into detail about your little spat but I want to know more about a certain arcane artifact you've brought to my show."

Ketrick nodded and motioned to Vivian. She drew a deep breath.

"Here we go." She said to her demon. She adjusted the drape on the phone booth size object and pushed it across the stage.

"My administrative assistant will bring the object out for everyone to see."

"What is it, Robert?" Sebastian asked.

"If you will join me we can reveal the contents together."

Sebastian followed him across the stage and together they pulled the silky drape away in one sharp move and tossed it over Vivian's head. Vivian swore and struggled to remove the drape.

"That's it." She told her demon. "She is history."

"Calm down." The bat demon said. "Our plans must be long in execution. Remember, Ketrick is but a tool. This is a minor setback. Let his infatuation with Sebastian blind him to your ambition."

"Yes." She sighed as she pulled the drape off of her face. "Patience."

The glass cabinet beneath housed a slender, rounded beam made of black wood. Its peak had been sharpened into a spike. Dark streaks covered the wood.

"Can you guess what this is?" Ketrick gestured with his bony hands. "Behold, Dr. Sebastian, one of the original stakes used by Vlad the Impaler."

"The Artifact Hunter does it again!" Sebastian shouted and the trademark sound of a whip underscored her words.

"Vlad the Impaler? My audience is well aware of the identity of Vlad. He was the inspiration for the modern legend of Count Dracula."

Ketrick popped a lock on the front latch. He opened the door and ran a finger along the edge of the wooden stake and came away with reddish, dark flakes on his fingers.

"I found this stake buried in the castle once used by Vlad. In the cellar, there were over a dozen of these objects wrapped in leather and frozen in a subterranean cave that never gets above freezing." Ketrick held his fingers to his nose and inhaled. "Ah, the fragrance of human blood."

Vivian rolled her eyes. The man was so dramatic! She hoped he didn't pop the dried blood into his mouth. That would reveal his fangs. And, such attention this early in her game might ruin everything.

"Robert, tell Dr. Sebastian how this stake was used by Vlad." Vivian said.

Ketrick glanced at her and blinked. He had forgotten she was there. "Oh course, Vivian. I was about to tell Dr. Sebastian more. Please don't interrupt us again." He turned his back on her and her face warmed with anger.

"These stakes would be erected outside Vlad's palace. He would take his victims and place them upon the stake. Slowly over hours and sometimes, days, gravity would press the stake through the body. It is said he caught blood from his victims in a golden chalice and drank it." Ketrick smiled. "He preferred his blood warm."

Vivian watched the color drain from Sebastian's face. "Just a minute, Robert. Are you telling my viewers this stake is covered in human blood from the 1400's?"

"Yes." Ketrick snapped his fingers and gestured to Vivian with an open hand. She hurried off the stage to retrieve the object he desired. She returned and handed Ketrick the golden chalice. Ketrick took the cup from her and held it up to the audience. The room filled with murmurs of fascination mixed with horror.

"Dr. Sebastian, I hold one of Vlad's chalices in my hands.

Notice the mark of the wolf dragon etched into the side of the chalice."

Vivian watched the naked lust fill the woman's eyes. Sebastian reached out a hand. "May I?"

"Of course." Ketrick handed her the chalice.

Sebastian held it up and let the studio lights catch the golden surface with reflected light. "Audience, you have no idea how fortunate you are tonight to see such a thing. The historical figure who inspired the fictional Count Dracula drank human blood from this very cup." She lowered the cup and walked toward the nearest camera. "Imagine an arcane perversion of the last supper. Assembled around a table set on the grassy hill beside Vlad's castle are his minions. Can you see them nervously gazing at their master as he ogles this cup? Vlad sends a servant scurrying to the nearest stake. He returns with this chalice filled with warm blood. Can you hear the screams of agony echoing around this ghastly tableau?" Sebastian gazed down into the chalice and the reflected studio lights illuminated her face in golden light. "Vlad lifts the cup to his lips and the warm blood trickles down his chin as he drinks."

Ketrick stepped forward and took the cup from her hands just as it reached her lips. "Probably not a good idea to put your lips to this cup."

A flicker of irritation crossed Sebastian's face. Vivian chuckled. Ketrick had spoiled Sebastian's dramatic moment.

"Vlad particularly enjoyed listening to the children of the night." Ketrick said.

"Who?"

"The wolves surrounding his castle. They feasted on the remains." Ketrick walked back to his cabinet and gestured with an empty hand. "Would you like to see the stake perform its magic?"

"What kind of magic?"

"Enough!"

Ketrick and Sebastian both whirled at the same time. They had forgotten about Anthony Cobalt sitting quietly on the sofa. He lurched to his feet and strode across the stage.

He shoved his face close to Ketrick's where only Vivian and Sebastian could hear.

"You are a fool, Ketrick. Dangling your goods in front of these mortals." He reached out and shoved the boom microphone away. "You are not a member of the Council for this very reason, Ketrick. And, I am now addressing your scorpion parasite, not the host. I will see that you are destroyed and sent to Tartarus where the master should have left you."

Ketrick chuckled. "Jealous, Cobalt? My power is greater than yours and you know it. Leave the magic to the big boys. Now, run along and fly your silly space ship." Ketrick paused and their gaze locked. "You know you will NEVER get out of orbit!"

Tension filled the air and Sebastian motioned to the nearby production booth. "Let's hear from one of our sponsors."

Sebastian shoved the two men apart. "I don't know what you ladies are fighting about. Frankly, I don't care. But, you are ruining my show and I will not stand for that. Cobalt, sit your butt back down on my sofa and let Mr. Ketrick finish his demonstration."

"You whining puke of a woman!" Cobalt hissed. "You will not talk to me that way. I have had enough of your silly theatrics. Whatever this charlatan has planned, you deserve it!" Cobalt stormed off the stage.

"Oh, let him go, Dr. Sebastian." Ketrick said. "He's only interested in publicity and getting more investors for his space station. But, I will guarantee you an unprecedented spike in ratings if you will allow me."

"Besides, that little argument just shot your ratings through the roof." Vivian said. Sebastian glared at her.

"I don't think I asked your opinion." She ran a hand through her hair and shouted. "I need a touch up! Now!"

Ketrick's gaze drifted beyond Vivian. She turned. Cobalt stood in the shadows just off the edge of the stage. For a moment, his appearance changed; hardened into dark, faceted crystals and then the man was back. He disappeared into the shadows.

"Who is that man?" Vivian asked.

"A rival." Ketrick licked his lips and shrugged. "Jealousy, Vivian."

An assistant touched up Sebastian's makeup while a second one probed at her hair. Sebastian glanced at the production booth. "I'm good! Let's get this show on the road."

The producer counted down over the speakers and Sebastian blossomed. "Welcome back to our little demonstration of an ancient and supposedly magical artifact. Before the break, our other guest, Anthony Cobalt became irritated at a spectacle that threatened to reveal the supernatural." Sebastian raised an eyebrow. "Science cannot even begin to touch upon the things that happen on the fringe of reality. Some people claim that physical objects carry with them a hidden power or ability. Claims are made that such talismans can be used to augment human abilities. I've been investigating these kinds of artifacts for years and I've never substantiated that any object contained hidden power. If there is power attached to these artifacts, it comes from some other supernatural source. And, of course, I don't believe in the bogey man! But, tonight, our guest, Robert Ketrick has promised to show us something truly magical." She turned away from the cameras and gave Ketrick a caustic look. "Right, Bob?"

"Of course." Ketrick grimaced. Vivian smiled. He hated being called Bob! He showed the cup to the audience tilting it so they could see it was empty. He turned it toward the camera. "Empty cup. Dr. Sebastian, thousands died upon these stakes and their blood soaked into the earth; into the soil; and the trees and plants drank of that blood so that it is not truly gone. For the power that lived and breathed in Vlad does not obey the laws of this universe. No, Dr. Sebastian, that power has gathered to itself every drop of that shed blood. It is here, trapped for us to see." Ketrick gestured toward the stake.

"Behold." Ketrick held the cup beneath a gnarled knob on the side of the stake. Blood begin to trickle from the knob until it filled the cup. Gasps came from the audience.

"Nice trick." Dr. Sebastian said. "Bob, I am the Artifact Hunter," and again the sound of the cracked whip filled the air. "And, I will never by fooled! I have debunked many charlatans in my day. I heard you were a magician, Ketrick. An illusionist."

"No, that would be number twelve." Ketrick whispered.

"What? Number twelve? What are you talking about?" Sebastian said. Like the cork of a champagne bottle, the knob popped away from the bark and blood gushed from the stake. Ketrick slammed the door to the cabinet and locked it. Blood splashed against the glass walls and pooled in the bottom of the cabinet. Vivian shook her head in dismay.

"What are you doing?" She grabbed Ketrick by the arm. "You're giving your true identity away."

Ketrick jerked his arm from her grasp. "You don't control me, witch!" His eyes were filled with fire and for a second, a dark spiral tattoo appeared around his eye. "I will not be upstaged by Cobalt and his dark angel." Ketrick grabbed the cabinet and spun it around allowing the blood to slosh inside. He shoved it toward the audience and the nearest camera. "Call me a charlatan? I will show you true vampire magic!"

As the audience rose to its feet, the blood filled the interior of the cabinet. Sebastian stumbled backward and fell into her seat. The camera operators kept filming Ketrick as he lifted the blood filled chalice for the audience to see.

"I make Vlad look like an amateur." He tossed back the blood in the chalice, swallowing the contents in a quick swallow. His lips dripped with blood.

Blood had now filled the entire cabinet and the glass walls shattered under the pressure. Blood flooded onto the stage and showered over the guests in the front rows. The air filled with screams of terror. They stumbled over each other as they stampeded from the studio. The camera operators abandoned their blood soaked instruments.

Ketrick laughed insanely and wiped the blood off of his lips with the back of his hand. He sloshed across the stage covered in copper scented blood and plopped down on the sofa beside Sebastian. He tilted the chalice and led one final drop of blood to fall on his tongue. He tossed it aside into the flood of crimson covering the stage.

"Robert!" Vivian waded through the blood. "That's enough!"

He glared at her. "Don't talk to me that way. You and your demon serve me. Me! Understand!" He reached with a bloody hand and took Sebastian's arm in his grasp. "Now, Dr. Cassandra Sebastian." He grinned exposing blood soaked teeth. "Or, can I call you Cassie? After all, you keep calling me Bob! Isn't that what your twin sister calls you? Cassie? Oh, look, I've stained your coat."

He giggled and wiped at her blazer, smearing blood all over it. "Now, Cassie, we need to talk about a certain artifact your sister is looking for. Oh, don't look so surprised. You can't fool me. One charlatan to another, I know you are keeping an eye on your sister's plans. I know you are watching her secret searches for a cure for her poor daughter. You know way too much and I need to be in the loop with you." He leaned into her and locked his gaze on her eyes. "Anthony Cobalt wants this artifact more than anything in the world. That is why he was here tonight. But, I need to get my hands on it before Anthony Cobalt to add to MY collection. And, you are going to help me."

"Vivian? Vivian?"

Vivian blinked and squinted into the bright sunlight streaming in through the limousine windows. Dr. Cassandra Sebastian waved her hand in front of Vivian's face.

"Are you having some kind of seizure?"

"No. Just remembering the details of our last meeting. Your show. The blood from Vlad's stake. Robert Ketrick. He wanted an artifact from you. An artifact that Anthony Cobalt was looking for." Vivian glanced out the window at the guard shack. "Area 613 belongs to Cobalt?"

"Yes. He moved all of his operations here months ago. This was once a military base. Not as famous as Area 51 so no secrets." Sebastian sat back. "So, you're here to take out this tenth demon? What was it Ketrick said? Something about number twelve? Is that some kind of code word? Or, are you dealing with the supernatural?" She paused and sipped more water. "There was a time when I didn't believe in magic, Vivian. But Ketrick's little appearance on my show changed all of that. We never aired that episode and we had to bribe everyone in the audience with a free trip to Las Vegas to convince them it was all a hoax."

"What do you want from Cobalt?" Vivian asked.

"I funded one of his satellites."

"Why a satellite?"

"The next advance in archeology." Sebastian pointed to the roof of the limo. "You'd be surprised what you can see from up there. Problem is, he's denying me access to it. Something about a deadline with an upcoming celestial event."

"You're spying on your sister." Vivian laughed and shook her head. "You're more like me than you want to admit."

Sebastian glared at her. "Okay, so we both want something with Cobalt. What are you going to do, kill him?"

"Of course not." Vivian said. "I just want to find out what he is up to. And, stop him."

"Orders from the Council?"

Vivian smiled. "You're good. Very good."

"My memory is good. Very good. He's this number ten, isn't he?"

"Get me into Area 613 so I can meet with Cobalt." Vivian said.

"And, what's in it for me?"

"You're looking for artifacts, right? In Dallas, there is a whole FBI warehouse of Ketrick's artifacts. And soon, they'll all come back into my possession. I know you're itching to get your hands on them."

Dr. Cassandra Sebastian leaned forward and extended her hand. "Deal. I just hope I'm not making a deal with the devil."

Chapter 9

Anthony Cobalt held the safety goggles in front of his eyes as the prototype fired a thin beam of orange particles across the chamber at the parabolic mirror. The particles swirled in a cloud of orange energy and contracted in a sudden pulse into a tiny pinpoint of light. The light coalesced into a small yellow jewel. For a second it hovered in the air and then fell into the hand of a lab assistant. Cobalt lowered the goggles.

"Very good, Dr. Barnard." He said. "When will this be ready?"

Barnard, a short, dumpy man in a beige lab coat lowered his safety goggles and his glasses were fogged. Beads of sweat glistened on his bare head. He smacked his lips as he thought. It irritated Cobalt to no end that the squatty man was so nervous and fidgety. Didn't he realize what he held in his hand? Energy compressed and stored in a remarkably stable jewel. Cobalt longed to rip out his throat but he needed the man's extraordinary brain to finish his plan. Barnard finally looked up at him.

"Sir, the satellite is already in orbit and we plan to deploy the mirror by the end of the week." He smacked his lips again and Cobalt watched the pinkish tip of the man's tongue dance around the interior of his mouth. Just one quick snatch and he could toss the tongue across the room.

"And once the mirror is deployed we can activate the energy beam?"

Barnard blinked behind his foggy glasses and smacked. "Yes sir. The energy will be compressed into storable particles."

"I'll hold you to it, Barnard. The sun is on a tight schedule and we don't miss to miss the Event. Besides, I'd hate to see what that instrument would do to a human being."

Barnard shuddered and glanced at him. "Sir?"

"Just a little joke, Barnard. You Americans are so stiff." Cobalt checked his watch. The thought of ripping out Barnard's tongue had made him hungry. His cell phone rang.

"Cobalt here."

"Sir, you have visitors." His secretary said. "Doctor Sebastian and a friend."

Cobalt frowned and glanced at Dr. Barnard. "The Artifact Hunter is a pain in my bum. Escort them to the executive suite and have lunch brought up. Tell my chef, I'll have beef tongue."

Anthony Cobalt stood at the window that overlooked the vast emptiness of the desert surrounding "Area 613". His main office building was over twenty stories tall and beneath him runways and gantries spread out in radial spokes. Cobalt Propulsion Laboratories covered over 4000 acres of desert. The buildings that radiated outward from the central tower reminded him of the drawings of a 1950s science fiction magazine he had collected as a child. Cobalt had built an empire in aviation and space technology. And, with Robert Ketrick, Rudolph Wulf, and Lynn Alba out of the way he had become one of the richest men in the world. There was only one woman who had more money than he.

"Anthony, it's been a long time."

He whirled to greet Dr. Sebastian and his eyes widened at the sight of the woman with her. So, it had begun. They had sent her after him. He forced a smile.

"Cassie, how good of you to visit." He took her hands in his and kissed both of them. "And, you brought a lovely visitor. Vivian Darbonne. Without the apostrophe."

Vivian held out her left hand. He pressed his lips to the back of her hand and gazed into her eyes. Yes. Her demons recoiled in terror. She knew he was indeed the host for the tenth demon. "Or is it Vivian Ketrick? Wulf? Did you marry Lynn Alba before she died in the tragic fire in Louisiana?"

"Vivian Ketrick will do." She smiled at him.

Cobalt straightened and motioned to the dining table. "Join me for lunch?"

"I'd love to, Anthony, but I need access to my satellite." Sebastian said.

Cobalt raised an eyebrow. Why did she need access now? What was she up to? "May I ask why?"

"You may ask." Sebastian smiled.

"Or, I can tell." Cobalt walked around the dining table to a far wall and waved his hand in front of a sensor. The wall sprang to life in a three dimensional hologram. Earth hovered along the floor and the vista looked out over space. Floating in midair was a golden bowl shaped object. "My space station, Vivian." He motioned to the station and it magnified. The ring around the outside of the station was shaped like a trapezoid in cross section and the center of the circle housed a series of spokes radiating toward a central crystalline structure. "The Eagle's nest. And here in far Earth orbit is my greatest triumph." He waved a hand and the station shrunk down to a miniscule size. A golden sphere appeared in the center of the image. "Soon, thanks to the control of OUR satellite, Cassie, this sphere will unfold like a space flower." He gestured. The sphere opened and unrolled into a huge, parabolic mirror of shimmering gold. "This solar mirror will focus sunlight onto the crystal at the center of my space station and I will produce an almost inexhaustible source of energy. I will change the world, ladies."

He motioned and the images faded back into the wall. "You see, Cassie, timing is of the essence and I need the satellite for an upcoming eruption of solar energy to test my new device."

"I need ten minutes, Anthony. Ten minutes on the satellite. And, my timing is critical."

Cobalt only suspected the real purpose of her need for the satellite. Perhaps it would benefit them both. "Very well.

After all, you did pay for most of the satellite construction." He took out his cell phone and dialed. "Barnard, Dr. Sebastian is here and she needs ten minutes of satellite time. I'll send her your way right after lunch."

"Now!" Sebastian interrupted him.

"Now, she says." He put the cell phone away and motioned toward the door. "By all means, take your ten minutes. If you are still interested in lunch, we will be here." But, he suspected she would be leaving in quite a hurry after her satellite surveillance. And, he would make sure her safeguards designed to erase her findings were bypassed. Sebastian nodded toward Vivian.

"Until next time, Vivian." She left and Vivian stepped out of the way of a group of servers rolling carts through the door left open by the departing Sebastian.

"Who's the monkey and who's the organ grinder, eh, Vivian?" Cobalt smiled and his hairless eyebrows arched. "Now, Vivian, please have a seat."

He pulled out a chair and Vivian slid into it. He sat opposite her and leaned back as the servers placed salad before them both. "Before you begin, I have a story to tell." He speared a bite of endive and placed it in his mouth. "Your precious Rudolph Wulf rescued someone from Communist Russia. The man was an expert on biosynthesis. He had developed a quasi-biological construct with the capacity of the human brain. Do you know how much information the average human brain can hold?"

"No." Vivian said. "I didn't come here to discuss Wulf."

"If the brain's storage of information takes place at a molecular level, one estimate of its capacity has placed it at about 3.6 X 10 to the 19th power bytes." Cobalt continued. "Wulf's apprentice had developed a cybernetic node capable of storing ten times the data of the human brain. And, this node had the capacity to connect directly to a human being."

Cobalt finished his salad. "I'm afraid I am not much of an herbivore, Vivian. I prefer meat. In South Africa, I routinely had what you would consider exotic game meat." The server took away their salad plates and placed a small bowl with a crenulated mound of pink flesh. "My favorite monkey was the golden monkey. Its fur was a lovely shade of red gold. Unfortunately, their brains are rather small. But tasty." He scooped a spoonful of the pink brain and savored its flavor. "I prefer it chilled but don't worry. It is poached. No parasites."

Vivian's face was pale and she pushed the bowl aside. "I don't do monkey."

"But, you have a liking for organ grinder blood, right? Vampires, Vivian. Really? But, you have moved on. I read about your dealings with Lynn Alba and number eleven. And, you came out of the whole affair a heroine." He finished the monkey brain and motioned to his server. The bowl disappeared. The server placed a plate of steaming meat before him. "Now, you will probably find this more palatable. Ankoli cattle meat is particularly sweet. They have these huge horns that can reach a length of two meters. And, although the horns make these cows appear dangerous, they are very docile."

He took a bite of the tongue and moaned with pleasure as he thought of Barnard's pink tongue. He chased the thought away. He had to focus on Vivian. "Now, back to our story. This man's development could have made me a trillionaire. But, the Council intervened and appropriated the device for their nefarious deeds. I believe you saw it recently on your visit to the Council."

"The Grimvox." Vivian looked like she had swallowed a snail.

"Even though the Council took my device, Dimitri owed me his life. Before the Council had him lobotomized, I made sure Dimitri installed a tiny chip ferrying all of the Grimvox recordings to me." He chewed another piece of tongue. My, how tasty! And, just the right amount of blood. "Of course, you won't tell them, will you? If you did, I'd have to destroy you." He placed his fork beside his plate and wiped his mouth with the silk napkin. "Oh, wait! I plan on destroying you anyway."

Vivian slid the plate aside. "You know why I'm here."

"Oh come now, Vivian. Don't let our little tiff keep you

from enjoying the best tongue this side of Paris!" He reached for her plate. "I'll finish it for you. My chief scientist made me so hungry for tongue this morning. I'm ravished." He placed another bite in his mouth. "You could, of course, have joined the Council's previous storage device."

"Joined?"

"An army of slaves driven to memorize the entire history of the Council. The Grimvox replaced them all. The Council could have plugged you into that wondrous organic circuit board. But, fortunately for you, they have the Grimvox." Cobalt finished the last bit of Vivian's tongue. "It took months to download the data from their brains into the Grimvox. Very messy." He wiped his mouth and placed the napkin beside his two empty plates. "Thanks to my secret access to the Grimvox, Vivian, my good woman, I know why you are here. They sent you to stop me. To kill me. And, they are keenly aware that you will most likely fail. Face it, love, they want to get rid of you. I am far more formidable than your last three foes. Ketrick? Really? The hubris of the man. And, his hair? Ghastly!"

Cobalt stood up and walked over to the windows. "And, Wulf? He was a good host but his demon was too blood thirsty. Never had the long view. Alba was strictly ambitious. You know that time has worn on our kind, Vivian. Imagine an eternity pressed between the pages of a book filled with nothing but the story of endless revenge and no respite from that desire. No hope. Give someone no hope and the only alternative is death. We cannot die so there is another alternative." He turned and studied her still sitting at the table. "Insanity. Eternal insanity. But, there are those of us who have managed to cling to shreds of our sanity, Vivian. We are those who comprise the Council."

He moved to the head of his table. "Tell me, Vivian. Is it getting more difficult to control them?"

"I don't know what you're talking about."

"Of course you don't, dear. Did you acquire one of the Vitreomancers? They are particularly kooky! All that vivisection and sucking out eyeballs." Cobalt shivered. "Such a base and macabre way of conducting the Master's business. I bet that last one has cowed the others, yes? It's taking over. And, if you are not careful, it will soon control you. And then, its insanity will leech into your mind." Cobalt sat down beside her. He leaned close to her face. "Ah, yes! There it is! I can see it in your eyes. Vivian, human insanity is nothing compared to the eternal insanity of the damned. Let me destroy you now and you can find painful rest with those in Tartarus."

Vivian bolted up from her chair and it fell over on its side. "Stop it!"

"Ah, now there is the vixen that may yet triumph over these cursed adversities. Good! Stoke that anger. Cling to the power of the human capacity to choose. We've lost it, you know. You are not yet truly damned, Vivian and the prospect of eluding your eventual fate will keep you sharp and in control. Remember that."

"Why are you telling me this?" Vivian hurried away from him and was stopped by the huge windows.

Cobalt laughed and joined her. "Because, my dear Vivian, I don't care what you do to me. Take my seat on the Council. You deserve it. Those fools can't see your power and potential. You took out three of the most powerful demons and you are keeping your seven demons under control. Quite a feat. You are an extraordinary woman."

He reached out and caressed her cheek. She recoiled from him. "Oh, don't be so coy. I know your type. I've dealt with hundreds of women just like you. If it were not for the fact I am no longer interested in having a tryst with anyone I would consider wooing you."

"Wooing? Really?"

"I can, at times, be a bit old fashioned. I was the second to be created after Lucifer. I am almost the oldest, Vivian." He turned and surveyed his kingdom. "You can have my seat. I have my own grand plans and soon, the Council will never be able to touch me."

Cobalt reached into his pocket. He handed her a yellow jewel no larger than a lemon seed. "You hold in your hand enough energy to power this entire complex for 24 hours.

Imagine the particulate energy I can now produce from the solar emissions."

"What is it called?"

"The Sunstone, of course."

"Do you realize what this will mean for the world's energy corporations?"

"Do you realize what it will mean for me? Not just riches, Vivian. But, incredible leverage. I will simply become the most powerful man on the planet. Once the Eagle's Nest is complete, I can begin harvesting vast energy from solar emissions from the parabolic disc you saw in the diagram."

"And this is why you no longer care about the Council?"

"I will control the world, Vivian. That is something even the Council could never achieve. And then, there is my other project."

He sensed Vivian stiffen beside him. Good! She had taken the bait.

"Other project?"

"Have you heard the tale of the first sighting of a UFO? Well, it took place at Mt. Rainier in 1947. I have that craft! I have the first sighted 'flying saucer'."

Vivian glanced down at her hand in sudden pain. She stiffened and collapsed onto the floor.

"I guess the tongue was too much?" Cobalt said.

June 24, 1947
Mt. Rainier, Washington

A long, long time ago, Tyhee Sahale became angry with the people and ordered a medicine man to take his bow and arrow and shoot into the cloud which hung low over Takhoma. The medicine man shot the arrow, and it stuck fast in the cloud. Then he shot another into the lower end of the first. He shot arrows until he had made a chain which reached from the cloud to the earth. The medicine man told his children to climb up the arrow trail. Then he told the good animals to climb up the arrow trail. Then the medicine man climbed up himself.

Just as he was climbing into the cloud, he looked back.

A long line of bad animals and snakes were also climbing up the arrow trail. Therefore, the medicine man broke the chain of arrows. Thus the snakes and bad animals fell down on the mountain side. Then at once it began to rain. It rained until all the land was flooded. Water reached even to the snow line of Takhoma. When all the bad animals and snakes were drowned, it stopped raining.

A Cowlitz Indian Legend

Kay Ball folded the document in half and stowed it in her backpack. Night waned outside her tent and the wind had abated. She had been unable to sleep with the anticipation of reaching the cave in the early morning hours. If the legend were true, then the cave might yield the Portal. Could this medicine man have passed through the Portal centuries before? It was the best lead she had found in years.

The presence within her quivered with anticipation and excitement. In the years since she had finished her secret work during World War II, she had researched the legends of North American Indians. After centuries of failure to find the Portal in other lands, Kay Ball had decided to turn her attention to North America.

The flaps on her tent moved and a dark face appeared against the pale dawn. "Madam, we have found the cave!" Sam, her Cowlitz Indian guide, had found the marks left in the mysterious earth by moonlight. Ball smiled and shrugged into her heavy coat.

"Let's see it!"

Sam led her and the other three members of the expedition up a long crevice in the upper stretches of the glacier and into the early morning sunlight. In the past two days, four members of her team had perished on the treacherous ice, but Ball had not relented. She had searched for the Portal for years and she was not about to give up. Sam gestured to a huge crack in the side of the mountain.

"The Mouth of God." He smiled. He had lost all of his teeth and his face was a mass of dark wrinkles beneath gray, stringy hair. "It is here the medicine man saw the cloud."

The opening in the mountain was huge, spreading from right to left for over one hundred yards. It turned down slightly at the ends as if God was frowning. Ball gestured to one of the other men.

"Mercer, go get the package."

Mercer, a small, wiry man was lucky to be alive. He had survived years of harsh work on the Manhattan Project as a lowly lab assistant and his appetite for money was immense. Ball had paid him dearly for the package. His incessant coughing had troubled her at first. But, the man suffered from radiation poison. He would never live to spend his money.

Sam led them into the interior of the cave. The roof stretched far above them and the walls were dark, crenulated rock. It had been decades since the last eruption but the cave was a huge lava tube. And, if her information was correct, at the back of the cave would be the fissure. Science had established much in the twentieth century and most scientists she had met during her stent with the Manhattan Project had balked at the supernatural. But, during the hours of flying them back and forth between their desert prison and their Washington meetings, Kay Ball had pried much information from them. Here, in this cave, was a strong source of radioactivity of unknown origin. And, the lava just beneath the surface would supply enormous power. Here, the magnetic field of the earth and the geothermal power of the volcano and the presence of radioactivity might approach the combination that could be responsible for the Portal. She could only hope.

Mercer trudged up the icy path behind her pulling the sled behind him. The package was about the size of a large suitcase and encased in bright aluminum. Ball nodded to Sam. "Take us to the fissure."

It took almost an hour to make their way into the depths of the cave over broken rims of lava and shattered shards of rock until they found the fissure. It stretched from the floor twenty feet into the air. Steam hissed from the insides of the fissure and Ball could see a faint, reddish glow. She pulled her magnetometer out of her backpack and studied the glowing dials. The magnetic field here was off the scale and oscillating wildly.

"This is it!" She shouted. "Place the package at the base of the fissure. Hurry!"

Mercer glared at her and pulled the sled across the broken rock. The two other men helped him wedge it into the base of the fissure. Ball pointed over her shoulder. "Sam, get them all back to the opening where you will be safe."

"What about you?" His dark eyes flashed in the glow of the fissure.

"I'll be fine." She whispered. Her inner power would protect her. If this worked and the fissure opened then she would have to hurry through the portal.

Mercer rushed past her as she dug the remote control out of her backpack. It was big and bulky with an antenna three feet long. She extended the antenna and waited until the men were out of the cave. It didn't take that long to run back to the opening when you weren't dragging a suitcase containing a small atomic bomb!

Kay Ball concentrated and the dimensions shifted around her. She reached out and opened a doorway into the heavenly dimensions. While profound and deep, they did not afford her inner power a passageway off this world. But, it would protect the woman from the blast. She regretted that the men would die. But, that was the price they would pay for believing her lies.

She closed her eyes and pressed the button on the remote. Light brighter than the creation event flooded around her, deflected by the extra-dimensional shield she had erected. Her supernatural tinkering with the bomb had insured that the blast would be mostly high energetic particles and not concussive. The energy would not collapse the cave but it would kill the men instantly. Already, her body glowed with the wispy edges of energy that were leaking through the shield. But, she would survive long enough.

The light died and in front of her, the fissure shivered and shimmered with mirror energy. Suddenly, the thing came through the mirrored surface. It was huge and a flat, silver circle of metal gleaming like a new moon. Flashing metal and lights, it came at her and turned horizontal at the last minute and barely missed decapitating her. Behind it, eight circular smaller objects came through the portal. The last scooped down and something grabbed her. She was jerked from the floor of the cave by an invisible force that snugged her up against the floor of the last object. She was glued to the bottom of the craft, her face aimed toward the floor of the cave. It rolled past quickly and she saw the ashy remains of Sam, Mercer, and her two assistants. They never knew what hit them.

The crafts pulled out into the clear afternoon air and began to circle the mountain. Kay Ball tried to curse. She had to get off of this craft and back to the Portal before it closed. These craft had come from another world; another area of space and she could care less of their origin or their abilities. The lead disk led them on a serpiginous flight around Mt. Rainier and through some of its valleys and around some of its lesser peaks. To an ordinary human this would have been exhilarating but Kay Ball was frustrated. She struggled against the force that bound her to the bottom of the craft to no avail. She couldn't even move her arms. The force was more than natural. She could not escape the body and return to the cave!

Far in the distance, she saw an aircraft. Even from this far away, she could tell it was a CallAir A-2 most likely on some kind of business trip. Could the pilot see the ships flashing and wheeling around her?

The crafts assumed a pattern of flight and moved in and out of the peaks around her. To her side, the airplane had moved into a parallel path. The pilot had noticed them. Maybe his presence would alert whoever or whatever was piloting these ships and make them return to the fissure. As if the occupants had read her mind, the ships headed away from Mt. Rainier toward the nearby Mt. Adams. No! They had to go back! But, she realized the lead disc was trying to elude the airplane. Their speed increased to an incredible rate and then the ships all turned at a steep angle, returning on the other side of Mt. Rainier.

The disc headed down toward the glacier and hugged the side of the mountain. The smaller ships followed and Kay Ball found her face just yards away from the icy slope of the mountain. Wind whipped around her and her face had long since grown numb. She was able to move her eyes and watched as short peaks and jagged boulders skimmed by just inches from her face. The ground dropped away and they were back in the cave. The ashes of her team had been blown away by the ships' passage. The large disc stopped and the smaller craft lined up behind it. At the far end of the cave, the fissure still shimmered with the mirrored light of another world.

The forces holding her to the ship's underside released. She fell ten feet to the rocky floor of the cave. Pain lanced across her chest and legs. She pushed up quickly from the floor and turned to the fissure. Already, the disc was turning perpendicular to the floor and heading toward the fissure.

Kay Ball tried to run after the disappearing ships and her legs ached. She stumbled over the rocks as pain ran up her right leg. She had twisted her ankle. The large disc hovered just outside the fissure and the smaller ships slid beneath it.

Kay Ball fell onto the rocky floor and cursed. She screamed and screamed until her voice was hoarse with anger. She unleashed her inner power and the ground trembled and the world tilted around her. The roof began to crumble and the mountain collapsed around her. The first rocks crushed her legs. The next one collapsed her chest and the tenth demon moved out of the woman's mind. As he slid through the rock like it was air, he heard the woman gasp in shock as her normality returned. But, it was short lived as the mountain buried her forever and her soul slid reluctantly into eternal darkness. And, somewhere, buried beneath tons of ice, rock, and snow was one of the ships. He would return for it someday.

Kenneth A. Arnold, an American businessman and pilot reported the sighting of strange flying craft near Mt. Rainier yesterday. Over the next few minutes, he observed several objects moving through the sky like a saucer skipped over the water. These "flying saucers" demonstrated deliberate movement and maneuvers only a trained pilot could accomplish. The objects disappeared in flight toward Mt. Adams. Mr. Arnold gives no explanation for the mysterious origin of these "flying saucers" but his attention to detail and careful observations lend credence to his claims.

Jackson Rimmer, freelance report at the briefing held by Mr. Arnold at the office of the East Oregonian in Pendleton, June 25, 1947.

"Vivian? Are you all right?"

Vivian blinked and warm spit dribbled down her chin. "I'm sorry! What happened?"

"You seemed to be out of it for a moment."

She gripped her hand tighter. He mustn't find out about the bloodstone. It had given her the vision of these strange flying craft. What did they have to do with Cobalt's plans? She released her grip on the sunstone and it rolled across the floor. "The Sunstone. It shocked me. I guess I fainted. Can you get me some water?"

Cobalt helped her to her feet and turned to the table for a glass of water. She twisted the ring so that the jewel was facing her palm. She took the glass of water with her other hand. Cobalt dropped the Sunstone into his pocket. "I thought I saw a flash of light before you fainted. I will have to chastise Dr. Barnard severely for this. He assured me the stones were stable."

"So, Dr. Cobalt, you mentioned a sighting at Mount Rainier? A flying saucer? Really? You have a flying saucer?"

"Oddly enough, Vivian, Kenneth Arnold coined the term because he thought the craft he saw looked like a saucer skipping across the surface of a pond. I assume he hurled his wife's good china across the water instead of trying to skip a rock. The media ran with the idea of a saucer flying through the air. Isn't it strange how these ideas form? A man tosses a saucer across the surface of a pond to watch it skip over the water and soon, the world is looking for saucer shaped craft flying through the skies bearing extraterrestrial life. The term 'flying saucer' was soon replaced by 'unidentified flying object'. UFO! Surely you don't accept that we are alone on this world? I believe the forces at work throughout the universe make life possible in many places. Estimates are that as many as 10,000 sentient races could exist here within our galaxy. I assure you the ships I am working on are not of this earth."

"You have more than one?" I asked.

"One large ship from Mt. Rainier and a much smaller ship from Roswell. Both similar. Both made by the same entities." Cobalt's cell phone rang and he listened intently and smiled. "Excellent. Have the pilot meet me in the hangar in thirty minutes." He placed the phone on the table before him. "Vivian, would you like to see a real flying saucer?"

"Solar mirrors and energy gems that will make you the most powerful man on the planet. And now, flying saucers? Just what are you up to, Anthony?"

Cobalt's lavender eyes glittered. "I have been planning this since the doors to Eden were shut. My plans are as old as mankind and they have not changed, Vivian. I'm getting very close to completing a most complicated plan that you would never begin to fathom. I am about to acquire one of three components I will need to complete my device."

"Device?"

Cobalt gestured to the vista outside his window. "Why do you think I built Area 613?"

"To sell tickets to wealthy want-to-be astronauts."

"Oh, so much more than that, Vivian." He took her by the elbow and steered her to the window. "Every dollar spent, every drop of sweat, every working hour has been for the purpose of preparing for a momentous event. Do you see that low lying mountain in the distance?"

"Yes." Vivian gently pulled her elbow from his grasp.

"It is Diablo Boca."

"The Devil's Mouth?"

"So it was named by the early Spanish explorers. Inside that ridge of stone lies a massive crater. We once believed a huge meteor fell in the desert eons ago. Now, we know the same tectonic activity that will one day produce the Yellowstone super volcano, flows beneath that crater. I use the geothermal power for all of Area 613. And, inside that crater, I have built the world's largest parabolic antenna."

Cobalt turned to study her with his lavender eyes. He arched his hairless eyebrows. "I have heard from another world, Vivian. It is out there somewhere waiting for mankind to discover. There is life in the stars! With the power of the Sunstones and my ships, one day I will stand on that world."

"And, what? Rule that world?" Vivian sneered. "Is that what this device will allow you to do? You disappoint me, Cobalt. You're nothing more than a demonic egomaniac with delusions of godhood."

"Delusions, no." Cobalt's eyes filled with fire. "Godhood, yes." He reached out and straightened her hair. "You are a stunning woman. But, I must tell you that frankly, the tenth demon and I can swallow your demons whole and leave you an empty husk."

Vivian stepped away from his hand. "Stop it!"

"Oh, don't worry. I will not allow you to communicate with the Council from this moment on unless it is on my terms. You will accompany me until my project is completed. And, if you agree to not stop me, when I disappear from this world you can claim victory and your seat on the Council."

"You want me to lie to the Council?"

"Of course! It is what we do. Our master is the father of lies." He stepped closer to her and his hypnotic eyes glowed with demonic power. "What harm can it do for me to complete my task? Now, do we have an agreement?"

"Of course, honey child." She said in her best Southern accent. "Now, how about you take little old me on a ride on a real flying saucer?"

Chapter 10

UFO cults share three interrelated beliefs: (1) Flying saucers are real. (2) People are in touch with alien intelligences associated with flying saucers. (3) The messages given by aliens are of immense importance to human beings.

Kenneth Samples in "Lights in the Sky and Little Green Men"

Vega opened her eyes and stared at the purplish light dancing across the ceiling. Was she in the ship with the weird music? She sat up on a low bed. Her feet hung off the end and she rubbed her head. It hurt. Polished copper formed the walls of her small room. A rounded doorway with beveled edges led out onto an empty corridor of the same metal.

"Hello, anyone?" She said. No answer. Was the Guardian here? Probably not. He had promised to rescue her and then fell into the river. She was alone and most likely in the hands of the Major.

But, what about the others? What should she do? The Guardian said he would find her. So, for now, she had to do what the Guardian had taught. First, reconnaissance. Get the lay of the land and determine if you are in enemy territory.

Vega stood up shakily and made her way to the door. She had to lean down to keep from hitting the top of the open doorway. She stepped into a strange corridor. The floor was flat but the walls were circular.

A strange buzzing, jittering sound came from her right. She rounded a gentle curve and gasped at the sight of the thing barring her way. The creature was as dark as night and seemed to vibrate, moving in and out of visibility. It had four wings that jittered and quivered with movement. The head spun and stopped, spun and stopped, each time showing a different face. One face looked vaguely human but deformed and squashed. Another face bore the crooked beak and steely eyes of an eagle. And the creature was not alone. Three others just like it stood to each side and behind so that each face pointed in all four directions. The four creatures had their backs to each other and one set of quivering wings touched the tips of the adjacent creature's wings forming an obscene quadrangle.

Vega shoved her fist in her mouth to keep from screaming as the creature walked toward her on hooved feet. The Guardian told her to keep quiet when danger neared. Be still and maybe danger will pass you. She stepped aside and pressed her back against the warm coppery wall as the creature quivered and jittered past her. The vibrating sounds disappeared as it moved around the curvature of the corridor.

Vega fought to control her breathing and kept moving down the hallway until she reached another open doorway. From her angle she could see one of the winged creatures standing before a holographic projection. Long, spindly arms came from beneath the wings. It had four hands! The thing touched the hologram and the image changed revealing an aerial view of the lights of cities and highways. They must be traveling far above the earth. Yes, the ship had picked her up and flown away from the swamp.

A shadow fell over a console just inside the doorway and a woman appeared. She was shorter than Vega and had no hair on her bare scalp. Strange brown spots covered most of the top of her head. She had high cheekbones and exotic eyes the color of tiger skin. The woman was dressed in some kind of Halloween costume all white and shiny with a high, peaked collar hugging her long neck. She sat before the console and a screen came to life and illuminated her face in multi colored specks. Vega could not see the screen.

"Master, I must speak to you immediately." The woman

said to the screen. Her voice sounded strange and hypnotic. "We have located all thirteen of the children."

A man's voice issued from a speaker before the woman. "Very well, Magan Celeste. Unfortunately, you set off several radar screens in northern Florida."

"I am sorry about that. It took your crew longer to figure out the stealth shielding than I thought it would. We were all quite surprised by the storm's effect on the school, you see." Magan Celeste frowned. "These helpers of yours are a bit disturbing. They are not totally reliable and I fear they will frighten the children. It would have been better to train some of my church members to fly this craft."

"Your congregation must oversee the workings at Diablo Boca, Celeste. Besides, only my helpers can fly this craft because of their unique capabilities and their connection with the pilot." The voice said.

Magan Celeste lifted a thin finger and pointed it at the screen. "Yes, I need to talk to you about that. Your original pilot refused to obey me so I had to make modifications. Besides, the pilot is more frightening to the children than your helpers."

"Perhaps that is why the craft was not hidden from view, Celeste. You will obey my orders without question, do you understand?"

Celeste raised an eyebrow and the brown spots on her scalp reddened. "We had an agreement. I gather the children and you allow my church members to worship them. I took you at your word."

"And, I need you to follow my orders to the letter, Magan Celeste. To the letter!"

"When can I take the children to the Church? The supplicants are eagerly awaiting the arrival of the Children of Anak."

"I have a visitor I would like to take for a ride in the smaller ship." The voice said. "Take your ship back to the compound and I will assure that the pilot cooperates."

Magan Celeste paused and glanced down at her console. "Oh, there is one curious thing, Master. When we were flying over the school, the instruments detected an unusual signature."

"Send me the data."

Magan Celeste flipped a switch. "Do you know what this is?"

Vega heard the Master gasp with excitement. "Can it be? How is this possible? It seems I'll have to extend my little test ride to recover this artifact. Call me when you arrive at the compound."

"Yes, sir." Magan Celeste swiveled in her chair and grimaced as she spoke to the creature at the other console. "Kromjaylik, set your flight path for the compound."

When the creature turned, Vega noticed its head did not spin. Kromjaylik's face remained stationary. A flattened snout and huge black eyes reminded her of a deformed cow. Its huge pink tongue raked over its nose and it saw her. It moaned a vibrating lowing sound and pointed one of its hands at her.

Magan Celeste stood up and stepped into the corridor. "Vega, isn't it?"

Vega turned to run but the four creature complex stood between her and the open corridor. Someone touched her arm.

"Vega, don't be frightened. I am Magan Celeste, Mother Superior of the Church of the Children of Anak. I mean you no harm."

Vega jerked her arm out of the woman's grasp. "What is this place? I want my mother."

Celeste nodded and smiled benignly. "Of course you do. But, Vega, I had to rescue you and your friends from the swamp after the storm drove you from your school. If I had not rescued you, then Major Miller would have found all of you."

Major Miller didn't have creatures like this woman did, but she feared him more than any other person on the planet. "The Guard--" She paused. No one must know about the Guardian. "Mother said she would never let Major Miller get his hands on us. She said he would," she swallowed. "Cut us up and experiment on us."

Celeste tilted her head and nodded. "Yes. All we want to do is to worship you."

"Worship?"

"Vega, you and your friends are very special. Surely your mother told you how special you are? Why do you think she has worked so hard to keep you safe and to keep you hidden from Major Miller? Once I realized you and your friends were in danger and that Major Miller was very close to capturing all of you, I used my Master's special flying craft to rescue all of you." She stepped closer and her hypnotic golden eyes captivated Vega in. "And, I will take you to a place of safety until you can all be reunited."

"What happened to our schoolmistress, Mrs. Donnelly?"

Celeste stepped back and looked away. "Vega, you must understand that I will be taking care of you now. Mrs. Donnelly is busy."

Vega felt nauseated and tried to swallow hard. "I don't trust you."

Celeste nodded. "Of course you don't. But, you need to know that I have a special connection with you and your friends. Your best friend is Altair, right?"

Vega pictured the tall, dark haired girl who was her best friend. "Yes."

"Well, I am her mother. So, you see, I have all of your best interests at heart, Vega."

"Then you'll let me talk to either my mother or Mrs. Donnelly?"

Celeste frowned and pointed behind her down the empty corridor. "Come with me, then."

Vega followed her down the corridor. She was at least a foot taller than Magan Celeste and she had to duck to avoid some crossing bulkheads in the corridors. The hallway curved away from her and Vega noted the other children from the school in their own cubicles. All of them were asleep. A doorway opened to her left away from the cubicles and she followed Magan Celeste into a large, ovoid room.

The room was encircled in consoles and view screens that revealed graphics and numbers. Others showed the interior of what appeared to be a hangar filled with white coated scientists bustling around the interior. In front of the consoles sat more of the winged creatures.

"Now, Vega, there is no need to fear. My little friends here are piloting this ship."

Vega looked down on the woman. "I want to see Mrs. Donnelly. Now!" Her voice thundered in the room and for a second, Magan Celeste's eyes filled with fear.

"Calm down, Vega, or I will have to give you something to make you rest. Mrs. Donnelly is right over here."

Magan Celeste motioned to the center of the room. A tall, crystalline column stretched from the floor to the ceiling. It's exterior was milky but with a wave of her hand over a console, the glass walls became transparent. Inside, Mrs. Donnelly floated in liquid. Her long golden hair surrounded her face like a veil. Her eyes were closed and her diaphanous gown floated in the liquid.

"Why is she in that water?" Vega reached forward and touched the glass. It was cold.

"She is receiving a, uh, treatment."

Vega nodded. "So, the healing waters are mending her wounds?"

Magan Celeste gently turned her away from the crystalline column. "Yes, the healing waters of Anak, my dear."

"Will Mrs. Donnelly be OK?"

"Why, yes, my dear."

Something dark passed down the corridor behind Magan Celeste. Shadows played over the walls beyond the opening and a sickly, greasy sensation crawled over her mind. Dread filled her heart and she glared at the doorway with fear. The music began, low and tremulous and it struck a numbing chord in her mind. She shook her head in fear. The air filled with the foul stench of rotten eggs and ozone. "What is that?"

Magan Celeste frowned. She turned toward the corridor and the shadowy substance solidified. Green eyes glowed within a distorted, deformed head. A mouth formed in the dark cloud of evil and a red tongue darted forward.

"I belonggggg in theeee tubeeee." It hissed. "She is an abominationnnnn. Take her ouuuut."

Vega's heart raced and she backed up against the cool glass tube behind her. This thing was evil. It was bad. It was everything bad she had ever thought of and more. Worse than the winged creatures. Magan Celeste planted her hands on her hips and shouted at the doorway.

"You stunted toe growth, get out of here! You'll ruin everything. She can power this ship far better than you did. Or, do you want me to tell the Master that his main pilot can't access the Sunstones for power? It was your fault we were picked up on the radar and you're frightening the children. No, go back to your cocoon or I will call the Master."

The thing in the doorway swayed and seemed to move from solid to cloud to solid. The eyes pulsed. "It is myyyyy job. Miiiinnnnneee. Howww did you getttt herrr to do thisssss?"

"I threatened to harm someone! Now, go!"

The eyes blinked and the dark form swirled into a cloud of darkness. It withdrew and only then did Vega realize the entire room had filled with shadows. Light returned. Sound returned and she heard the gentle rustle of Magan Celeste's gown and the beeping and keening of the instruments. The winged creatures had frozen at the thing's appearance and now returned to their instruments. Mrs. Donnelly's voice echoed in her mind.

"Vega, go back to your cubicle. I am not harmed and I know what I am doing. I am watching over you."

Vega whirled and for a second saw a twitch in one of Mrs. Donnelly's eyes. The reflection of Magan Celeste eclipsed Mrs. Donnelly.

"Now, if you're satisfied that Mrs. Donnelly is fine, I need you to return to your cubicle until I can take all of you to the hangar. You will finally be safe." Magan Celeste smiled as if the entire world seemed right.

Chapter 11

Since the dawn of mankind, we have looked to the sky for our salvation. We long to reach out to other worlds, to other beings far superior and more powerful than we pitiful mortals. Humanity's only hope lies in the stars. That is why I have worked so hard for decades to perfect the first privately funded orbital space station. And, today, we celebrate my first orbiter filled with civilians longing to touch the stars. This is our first step to moving beyond this world to the next.

Anthony Cobalt, news conference on the day the Cobalt Aquila reached orbit.

Cobalt leaned forward and allowed a blue light to play across his right eye. "Retinal screening." He said as the metal door to the hangar opened. The room beyond was immense and as large as a domed football stadium. The sight of the huge ship sitting at one end of the building took her breath away. It was saucer shaped with a slightly upswept underside. From the periphery of the ship, the upper surface sloped gently until it met a dome that occupied the center of the ship. The upper surface glowed a faint, luminescent green and the undersurface glowed with a lavender light. The air tingled with a low level of static electricity and she smelled ozone.

"Flying saucers? You weren't kidding!" Vivian said. "You have a working flying saucer?"

"Have you ever heard of Ralston Mead?"

Vivian glanced at his lavender eyes. "No."

"Pity. He was a most fortuitous man. Ralston Mead made a small fortune writing science fiction stories in the 1950's. Hollywood loved his pulp fiction style and several of his stories were converted to alien monster movies of the time. Ralston was smart and insisted on a percentage of the movie gross instead of one lump sum for his stories." Cobalt reached out and ran a hand along the edge of the large ship's surface. "Unfortunately, Ralston loved to gamble and ran afoul of the mobsters in Vegas. So, he hid away in the mountains of the northwest prospecting for gold. One day, he discovered a cave on the slopes of Mount Rainier. And, inside that cave, he saw this gleaming metal. He thought he had found something more precious than platinum."

Cobalt smiled at her. "Fortunately for me, my partner, so to speak, found him and soon, this ship was unearthed in all of its glory. It is from another world, not just a pile of precious metal. I secreted it away to Diablo Boca. That was long before I entered the picture with my fortune gained from the aviation industry. Would you like to meet Ralston Mead?"

Vivian shook her head in confusion. "He is still alive?"

"In a sense." Cobalt touched the metal surface again and the purplish glow intensified. Something dark formed on the surface, a murky plum hued stain that coalesced into the figure of man. A bearded, aging man seemed to be trapped beneath the surface of the metal like a bird slammed against a windshield. His eyes betrayed his horror. "Let's just say Ralston gave his all to our project. You see, he tried to betray my partner and sought help from the government. My prior host stopped him before beginning his own new partnership with number ten."

The image of Ralston Mead faded from view. Who else was trapped in the strange metallic surface of this thing? For a second, she saw the ghostly figure of Kay Ball. She looked away. "So, the government knows about these ships?"

"My dear Vivian," Cobalt placed a hand on her shoulder and turned her to face him. "The government has been working with a 'flying saucer' since the 1940's. They finally abandoned the work when their efforts proved fruitless in understanding the propulsion system of the ship recovered from Roswell. However, they learned enough to develop the stealth bombers at Area 51. But, this original craft was kept secret by my predecessor until I came along with my new propulsion system powered by the Sunstone. By the time the military knew where I was headed with my research, they had sold me everything and signed over complete control of the ship discovered at Roswell. I keep my secrets, Vivian. They never knew of this ship. Only a few chosen employees know of this hangar. What you see before you is the oldest ship."

"You have more than one?"

"This one and a version from Roswell currently in use." Cobalt's eyes gleamed with excitement and Vivian realized the color of his eyes matched the lavender glow of the large craft.

"I'm dying to ask, honey." Vivian stepped closer to him and placed her hands on his chest. "Have you been to Mars yet?"

Cobalt frowned. "Traveling to the stars is not possible. Yet. My ship's abilities are confined to the Earth's orbit."

Vivian watched a dozen men in white jumpsuits scurrying around the periphery of the ship. She walked slowly around it marveling at the streamlined metal. But, as she tried to focus on the surface of the ship, it seemed to blur. It was as if the ship was not totally within their reality. Something about the surface disrupted her human senses.

"Don't you sense it?" The Vitreo demon hissed. "Feel it? The power? The purity of distilled evil?"

Vivian paused and reached out with her demonically enhanced senses. Waves of evil energy washed over her. Her lesser demons appeared, reveling in the evil. The Vitreo demon swelled with power threatening to eclipse Vivian's will power. "Stop!" She hissed out loud.

"They love it, don't they?" Cobalt whispered in her ear. Vivian jumped and fell up against the surface of the craft. The skin vibrated against her and her demons rejoiced. Instantly, she heard the screams of the dead. How many had died to service this ship? Were their souls trapped in this obscene metal?

She pushed away from the velvety surface and ran across the hangar holding her head in her hands.

Cobalt followed her. "They sense it! They know my plans, Vivian. The evil this world has seen since the dawn of time is nothing compared to what I will bring upon this universe!"

"Stop!" Vivian shouted and she slid down against the wall. "Stop it! All of you!" The Vitreo demon unleashed a verbal barrage of unintelligible speech at the rest of her demons. The chatter in her mind quieted and she realized her face was stained with tears. "What have you done, Cobalt?"

"It's not what I have done, my dear. It is what I will do!" He laughed out loud and Vivian fought against the nausea and dizziness. Cobalt glared at her. "Now, get your puny demons under control so I can take you for a ride. If you act up again, you'll join Ralston Mead."

PART 3

ABDUCTION

Close encounters of the fourth kind (CE-4): Direct contact occurs between alien beings and the witness. Apparently taken aboard a spacecraft, the abducted witness interacts extensively with the spacecraft's occupants.

Close encounters of the fifth kind (CE-5): The observer suffers permanent physical injuries or death.

J. Allen Hynek
Center for UFO Studies

Chapter 12

Cephas Lawrence glanced at his wristwatch. It was almost 1 A.M. and still no word from Jonathan Steel. He would not sleep until he heard from the man. He puffed on his pipe, a vice from his past that he had resurrected in the days since their encounter with the eleventh demon. The thought of Rachel, the daughter of his one true love, Molly being in the possession of that fiend still haunted his dreams. Best then that he did not sleep. He heard the deck creak behind him.

Theo moved beside him like a huge shadow. "Papaw, Chief want to talk to you." He held out a cell phone.

Cephas took it and pressed it against his ear. Theo settled into a deck chair beside him and it whimpered with the man's weight.

"Jonathan? Did you find her?"

"No, Cephas. But, I found someone else. Someone from my past. I'm following him through the woods now. He may lead me to Renee's daughter."

Cephas bit down harder on the pipe. "Be careful, Jonathan. I have a very bad feeling about this. The good Major was involved in something that smacked of the supernatural."

"I found something at the school. It's a red jewel on a chain. Shaped like a tear drop or a drop of blood."

The pipe fell out of Cephas' mouth and thumped on the deck. He stood up and stumbled toward the deck railing.

"You okay, Papaw?" Theo asked.

"Yes. Jonathan, you said it is shaped like a drop of blood? Are you sure?"

"Hey, I'll text you a picture of it."

Cephas' mouth went dry and his pulse quickened as he waited for the screen of his phone to show the picture. An image of a red jewel nestled in the palm of Jonathan's hand arrived on the screen. It was shaped like a drop of red substance. He rubbed his mustache. "Oh, my! Jonathan, remember the Demon Rose, the symbols for the twelve demons on the Council of Darkness?"

"Yes. Spiral for number thirteen. Wolf dragon for number twelve."

"Chimera for number eleven." Cephas said. "And, Jonathan, a drop of glowing blood for ten."

Steel gasped. "No, no, no! Not now. No, it can't be happening."

Cephas wiped his mouth and Theo handed him his extinguished pipe. In his mind, he was back in the basement of Robert Ketrick's house, standing before an empty wall. Behind it he had found the thing. And that thing had a dark jewel just like the one Steel described imbedded in its chest. He had told no one for he was certain it would lead him to Molly. "Jonathan, we can't let the tenth demon get his hands on these children."

"Do you think Miller is the tenth demon?"

"No. There is another person working in the background, staying in the shadows." Cephas wiped sweat from brow. "He was pulling Miller's strings when Miller was looking for the Elixir of Anak. That person must be the tenth demon."

"I remember something from what the dog said."

Cephas raised an eyebrow. "The dog you rescued? You did not tell me it talked to you!"

"Mentioned the sons of Anak. And, said it was time to play. This is all connected. Why is this person pushing Miller to find the children?" Steel said.

"The children are special, that is all I know. And, the tenth demon will abuse them for his purposes. Forget the man you're following. Come back to the beach house now."

"No. This man is the only lead to the children, Cephas. I have to find him. Tonight." Steel said.

"Who is this man?"

Steel was silent for a moment. "Arthur Knight. He's alive, Cephas."

Cephas almost dropped his pipe again. He tried to swallow but his throat was dry. "Are you sure?"

"Yes. Where's Theo?"

"He's here."

"Let me talk to him."

Cephas handed Theo the cell phone and his gaze wandered out over the ocean. The stars were gleaming in the dark sky like jewels; like the red jewel that was in Jonathan's possession. How had Steel end up with the Bloodstone?

"Okay, Chief." Theo ended the call. "I don't like him being out there by himself, but he says for me to protect the two of you. He told me to take the two of you back to Shreveport. We're leaving first thing in the morning."

Cephas thought of the thing in the wall of the basement. Yes, it was best if he was there. Perhaps this thing could protect them.

The ship's observation lounge allowed Vivian to watch the earth stream beneath them at unbelievable speed. The lounge was on the edge of the ship and as far away from the source of evil as possible. Cobalt entered the lounge. He had changed into a one-piece pale green jumpsuit.

"Feeling stronger?"

Vivian sat in one of plush chairs attached to the smooth, metal floors. "What happened to me?"

"I once spent a winter in the forests of East Germany. I was only 21 but a young woman living in a farm house allowed me to stay for the harsh winter months if I would work for her. Mostly odd jobs. It was right before the wall came down and her husband had been dispatched by the KGB for spying. She was most accommodating until she found out I was also working as a double agent for the KGB. Didn't matter whose side I was on. My affiliation with that band of brigands was enough. Unfortunately, I was unaware she had discovered my secret."

Cobalt settled into a chair next to Vivian. "You see I was an outlaw from both sides of the Iron Curtain. I had left my South Africa behind and become a Communist to fight against apartheid. Or, so I thought. Turns out Communism was as tyrannical as apartheid so I switched sides and worked for West German intelligence until they learned about the unfortunate deaths of a carnival full of Europeans slaughtered in their sleep. I was only testing a new weapon that would drain energy from a living body and produce a jewel. I had to attempt many trials before I got it right."

"Anyway, the young woman learned of my identity and planned on betraying me to the local authorities. But, not before she plied me with poison from a certain root herb known to produce violent, painful death. And, here, Vivian is the point of my story. I filled the fireplace with wood and told her I would be back shortly after a trip to the market. She had planned on slipping the poison into a pie and needed flour. While I was gone, she would prepare the poisoned filling for the pie and after my untimely death she would collect the bounty from the local authorities. I lit the fire and when I came back she had succumbed to the poisonous smoke of the same roots burning in the fireplace from which she had extracted the poison. The source of her warmth was the source of her death. Evil repaid for evil."

"What's your point?" Vivian whispered.

Cobalt's smooth features gazed back at her from his reflection on the window. "My propulsion system is dangerous to you, Vivian. For your demons it is like blood in the water for sharks. If you were to stand in my control room, your demons would go into a feeding frenzy and you would not survive. That which empowers you will lead to your destruction."

"Your helpers? What are they?" She asked. Four of the jittery fly like creatures were stationed just outside the door of the lounge.

"Messengers of darkness. You need ask no more."

"Very well, where are we going?"

"To pick up a little something for which I have been searching for a long time. And, as you can see, we have stopped. My helpers will have located the person in possession of my prize and will have brought them on board. Shall we go to the cargo hold?"

Chapter 13

The aliens sometimes subject the abductee to what is described as a crude and painful medical examination. The aliens place the abductee on a table and remove his or her clothes. Often they then clean the abductee's body and examine him or her by hand or with a machine.

Kenneth Samples

Steel slid his phone into his shirt pocket and waited. He had tracked Knight through the woods to a large culvert that passed beneath a highway. Knight disappeared inside and Steel hurried across the road to wait for him to emerge from the other side. Steel leaned against a palm tree and looked down at the stream that emerged from the culvert.

A shadow stirred at the mouth of the culvert and Knight stepped into view. Before Steel could move, light gushed down from above and caught the man in its embrace. Arthur Knight stood transfixed as the music began. A sickeningly sweet odor filled the air along with the ethereal and pulsing music.

Steel tried to run but his legs were useless. Paralysis crept up his legs! He grabbed the trunk of the tree to stay upright. Below him, Knight seemed frozen in mid stride. Jonathan fought against the nausea and the weakness. The sweet smell of death gripped him. He glanced up along the cone of light and something hung against the stars.

A hole opened in the bottom of the craft and things

dropped out borne on the air by four wings. Against the blinding light, they seemed to blur and vibrate in and out of sight. Their spindly limbs stretched away in dark, reticular shadows along the streambed as they converged on Arthur Knight. Pulse pounding, scream inducing terror filled his mind. He tried to fight it and he clung to the tree trunk as if the sheer power of those hideous creatures would suck the very life out of him.

The things surrounded Knight. They joined hands and the four of them formed a square with the tips of their wings. The music's pulsations quickened and Knight levitated into the air, frozen with his eyes wide open. Along with the creatures, he ascended toward the craft above and disappeared into the interior of the craft.

The terror lessened and Steel tried with every thought to fight the paralysis but the music was too strong. His grip on the tree weakened and gravity triumphed. He toppled sideways toward the stream and bounced against the edge of the culvert. He rolled like a dead log down the slope and fell face down into the water. His eyes were wide open as his face plunged into the cold water and he held his breath, the only control he still had. If he could not move soon, he would drown.

His face pulled away from the water as his body floated upward. He gasped for breath and looked down at the sight of the horrific alien things surrounding him. The creatures' wings vibrated and jittered. Their individual heads spun and blurred in and out of focus. Even with their backs to him, the heads still paused and glared at him with demonic eyes filled with fire and evil and hatred. As he floated through the open iris his phone slid out of his shirt pocket and fell into the creek.

Vivian watched the bodies of the two men float upward into the huge cargo hold of the ship. The first man was dressed in dark clothing and his face was partially covered with black paint. A canvas bag dangled from his neck.

"Two men?" She asked.

"Unfortunately, the situation has been complicated by the presence of another man nearby who saw the ship."

A second figure appeared in the opening of the cargo hold and floated over to the waiting aliens. Vivian gasped as his face rotated toward her and his turquoise eyes gleamed in the purple light. She ran around the closing iris to the group of creatures. They were placing the inert body of Jonathan Steel on a metal table. She leaned over his open eyes and waved a hand in front of them. He was unresponsive.

"No worry, Vivian. He is asleep." Cobalt turned to the creatures. "Take them to the examination chamber."

"Baby, what am I to do with you?" Vivian whispered. The Council had warned her about harming Jonathan Steel. Did that include harm caused by Cobalt?

"No worry, Vivian." Cobalt leaned over Steel's body. "He is hypnotized by my unearthly music. We can dispose of both of them once I have my prize."

The smell hit him first, a mixture of stale sweat and a pungent chemical. Steel tried to turn his head against the leather restraint across his brow. Shadows moved at the periphery of his vision. He was on a metal table and just a few feet to his right Knight lay stretched out on a similar table tied down by leather restraints at his ankles and wrists.

Above the tables, darkness shrouded the ceiling. And, it was from this darkness that the abomination descended. The first hint of what hid in those shadows came with the sound of an odd mixture of bony clacking and soft, moist squeaking. As the thing moved into the lights mounted at the corners of the table, Steel wished he were stricken with blindness. Hell wore flesh! The thing could not have possibly existed in the real world. Yet it pulsated and writhed as it moved toward Knight's body.

It was half flesh and half metal. Appendages emerged from a huge blob of purple tissue vaguely human and resembling an arm here and a leg there. At the end of these gyrating appendages, metal tubes and needles and probes reflected in the light. Some of the appendages bore eyes that spun and angled to direct the placement of the appendages. Here and there a mouth like orifice burbled and murmured nonsensical sounds. The stench was unbearable.

Knight struggled against his restraints as the thing hovered above his bed. One huge head like knob on a metal tentacle thrust itself into Knight's face. It bore a host of bleary eyes. A huge, serrated mouth opened and a long pink tongue caressed Knight's face. Steel's heart pounded with fear and sweat poured from his forehead. What was this thing? How, in God's name could it exist?

From the shadows of the chamber the winged creatures appeared. They vibrated and gyrated and jittered with unearthly movements. Each head spun and froze then spun again. Four wings beat against the air producing a dry, rattling sound. They moved like flies that jerked and spun and jerked and spun.

"Nightflies." Steel whispered. "That's what you are, nightmare flies."

The creatures gathered at each corner of Knight's table and extended their wings to touch tip to tip. Through the vibrating diaphanous wings of the nightflies, Steel watched the thing from the ceiling hover over Knight.

A maroon appendage appeared with fine pincers like the feelers of a sea crustacean. The pincers tugged at Knight's shirt and the monstrous head swiveled to direct the pincers' progress. Knight's eyes were wide with terror as he watched another appendage with a tiny circular saw follow the pincer. The saw spun and hummed as it deftly cut away the shirt. The pincers pulled the material away. Other pincers descended and in seconds, Knight's shirt was reduced to tatters. A shallow metal box appeared and the tatters were placed in the box. An oblong monitor appeared in the thing above the table. Like some arcane X-ray, the tatters illuminated on the monitor. The nightflies milled about and murmured among themselves in a language Steel did not comprehend. The contents of the metal box flared up in bright flame and were gone.

The head descended toward Knight's bare chest and the

tongue left a slimy trail along his collarbone. The mouth whispered.

"I can taste its remnants on your flesh. Where is it?" The thing hissed.

Knight tried to speak but his mouth was paralyzed. The head turned toward the rest of Knight's body. The pincers slid down his legs and tore away his pants, his belt, his shoes and last, his socks. Each piece was cut, examined and incinerated until Knight lay in only his underwear, unprotected on the metal table.

The head appendage looked up into the morass of its body and a knobby appendage bristling with needles and pointed metal probes descended. The knob paused above Knight's arm and the needles shot forward into his flesh. Knight's mouth could not open but his moans of pain worsened as the needles and probes raked across his chest.

Another long and spindly appendage jointed like a deformed crab leg extruded from the creature. At the end of the appendage, a cluster of thin black tentacles emerged and writhed above Knight's face. His eyes widened even further in terror as two of the tentacles snaked their way into each nostril. Knight gagged and his body convulsed on the table. Steel had to look away. Even still, the shadow of the moving limbs played across his vision. Finally, the thing stilled. Steel slowly moved his head back to see Knight. The man's eyes were closed and his chest and abdomen were covered with tiny lacerations and bubbles of blood. His chest still moved. The tentacles withdrew from his nostrils with a slurping sound, dripping mucous and blood.

This thing was beyond horror. It was beyond evil. It was unimaginable that a God that loved him would allow such a thing to even exist. Had God abandoned him?

The nightflies detached their wing tips and came together between their two tables and conversed among themselves in an obscene language. Silence fell and as one, they turned toward Steel. In unity, they moved and took their places at the corners of his table. It was his turn.

The monstrous probe floated away from Knight's bed and

stopped above Steel's table. He tried to swallow and couldn't even accomplish that. His head and neck muscles were paralyzed. All he could move were his eyes. The hideous head thing descended out of the dark shadows and paused just in front of his face. He smelled its foul odor and its cold, exhaled breath chilled his skin.

"Where is it?" The mouth said. Steel wished he could speak. He wished he could scream at the thing. And, in a flash, he realized why he was paralyzed. Not only did it keep him from moving from the table, but it kept him from speaking. For, if he could speak he would scream the name of his Lord and send this thing back to the hell from which it had come. It had been long since he had contemplated that name. His anger at being a "reluctant draftee" in the battle against evil had eclipsed his connection with the Divine. As long as he dwelled in the presence of his Savior, he would be immune from this evil.

Steel realized in a cold wave of relief that this thing could not kill him. This thing could not own him. This thing was of the enemy and all it could do was frighten him. He caught his fear like a bird fluttering in a room and spoke to it as he calmed it in his hands. He told it to be still. He told it to be quiet. He told it God was in control and all would work out. He spoke the name of God and his heart rate lessened. He closed his eyes and realized he would not die this day. He would live. And, he would see that this abomination was destroyed.

The pincers descended and tugged at his tee shirt. The faint whiz of air touched his chest as the saw cut away his shirt. The tattered cloth dropped into the examination box. If it took his jeans, it would find the stone. Was that the object of its searching? The jointed arm with the tentacles descended from the darkness and writhed before his eyes. Steel glared at it; pouring every ounce of defiance and fury into his gaze. And, the tentacles paused, hanging motionless above his face. A drop of mucous fell and hit his cheek, ran down his face like clotted sweat and unexpectedly, Steel was somewhere else.

"Something in your eye?"

He glanced at the soldier crouching in the bushes beside him and blinked again. He wasn't used to wearing contacts. But, he knew his eyes would give him away so he had put on brown contacts. He wiped the tear from his cheek.

"Just dust. What's the mission again?"

The soldier shifted his rifle from one hand to the other and tugged his night vision goggles down over his eyes. His face was now hidden beneath the dark makeup on his face and his hood. "The Major will tell us when he shows up. Just be ready."

"Cut the chatter." Someone whispered behind them. He glanced over his shoulder at the other six soldiers hidden in the underbrush along the mountainside. The quarter moon gave enough light to make out their shapes huddled behind him. When he had learned of this mission, he had followed one of the soldiers. He had taken out that soldier and replaced him. Since these soldiers were all independent contractors, they didn't know each other. With the soldier's equipment in his hands, he was as good as the man's twin.

"What's your call sign again?"

He looked at the soldier next to him. "Hot Steel."

The soldier snickered. "Mine's Sawbones."

A shadow loomed over them. Hot Steel looked up into the eyes of Major Miller. "I thought I told you two to cut the chatter."

Sawbones shrugged. "Sorry, Major. Just getting to know my right hand man."

Miller was almost invisible in dark clothing. "The two of you are paired up then. The rest of you pair up. The cabin is just up the mountainside, a warm up shack for the ski patrol. I want a pair to approach from all four compass points." Miller poked Sawbones in the side. "You two take the rear entrance. Wait for my command and we all go in at once."

"We take the package alive. Each of you has six rounds of tranq darts." Miller crouched down and looked over Hot Steel's shoulder down the mountainside at the rest of the men. "Now, listen up. You were chosen because each of you is good at what

you do. I don't know you and I don't want to know you. We take out the target and then you scatter and rendezvous for the next target. Understand? Just know that the target is small, about four foot seven inches. He's wearing a red jumpsuit. But, don't let his size fool you. He can neutralize you in a heartbeat. Once you get a bead on him, hit him with the darts. It will probably take two or three."

Hot Steel raised an eyebrow. One dart was enough to take out a six foot, three hundred pound man. But, three? Miller stood up and touched his hand to an earpiece.

"Yes, Captain. We're in place and ready. I'll inform you the minute Neph 3 is in custody." Miller said. Hot Steel blanched. The Captain? His father was involved with this operation just as he suspected. Miller whirled and pointed up the mountainside.

"Men, go to it." He shouted in a loud whisper.

Hot Steel followed Sawbones up the rocky slope. The cabin was just ahead on the crest of the mountainside overlooking a ski slope. It was mid July and the slopes were covered with high grass and flowers. Sawbones motioned to his right and Hot Steel followed him up and around a large clump of boulders.

"Now, we wait for the signal." Sawbones whispered. The cabin was about twenty feet away and Hot Steel could see the rear door in the meager moonlight. Something flashed across the way and he saw two men take up their place to the right. "Who do you work for?"

Hot Steel looked into the man's faceless goggles. "Myself."

Sawbones chuckled. "Good answer. You're not one of the Major's men."

"Guess not. Are you?"

Sawbones spat into the bushes and shook his head. "I work for the man with the most money. There's the signal. Let's go."

Sawbones ran full strength up the slope and rammed into the back door without stopping. Hot Steel hurried after and brought his rifle up as he stepped into the interior. The cabin was one large open room and as he stepped through the open door, four soldiers burst through the side doors. Two more crashed through the front window. They stood in a circle looking at each other.

Huddled on the floor in a sleeping bag was a boy. He was asleep and he slowly stirred. Sawbones leveled his rifle and hesitated.

"He's a kid."

The boy looked much younger in the face than his size. In fact, he seemed like an oversized toddler. His eyes snapped open and he sat up with his right thumb shoved into his mouth. He blinked huge eyes as he looked at the men around him. Sawbones pushed up his night goggles and glanced once at Hot Steel. He shrugged.

"I think one dart will do." He raised his rifle and the boy casually pulled his thumb out of his mouth. He gestured toward Sawbones and the man hurtled through the air across the cabin. He crashed through the sidewall and splinters and wood geysered into the air. Hot Steel stumbled back. The other soldiers raised their rifles. The boy gestured in their direction. Two soldiers went straight up and through the roof. Moonlight gushed in through the hole and caught the boy in ghostly light. He slowly stood up and his face was twisted in anger.

"You bad men." He said and flicked his hand toward the other soldiers. They flew through the air and out through the open side doors. Hot Steel didn't know what to do. He was stunned. He raised his rifle and pointed it at the boy. The boy slowly turned his huge eyes toward him.

"You not a bad man." The boy said. He gestured toward Hot Steel and nothing happened. Hot Steel didn't hesitate and pulled the trigger in three quick successive bursts. The darts landed squarely in the boy's chest puckering his red jumpsuit. The boy looked down at the darts and his eyes rolled back in his head. He fell backwards onto the sleeping bag.

Major Miller ran into the room, his rifle at the ready. "I told the Captain this wouldn't be easy. Report, soldier."

Hot Steel walked over and looked down at the sleeping boy four times the normal size for a toddler. "It was Sawbones, Major. He brought down the boy."

Miller nodded. "His last act."

"What?"

Miller gestured toward the back door. "You're the only survivor, soldier. Get to the rendezvous point. Now!"

Hot Steel backed toward the door, his eyes drawn to the young boy. Miller touched his earpiece and Hot Steel heard a familiar voice over the speaker. "Report, Major."

"It's done, Captain. Lost seven men."

Hot Steel paused outside the door and stared at the body of Sawbones impaled on the metal frame of a ski rack. He reached up and pulled off the man's mask and looked at his lifeless face. He wanted to remember it for a long time. Suppressing the shaking of his hands, he reached down deep and found his anger and hatred for his father. He immersed himself in the feelings and found purpose. He headed off down the mountain slope to the next rendezvous. Sooner or later he would find his father and the man would answer for this atrocity.

Steel opened his eyes to bright light. Somewhere in his past he had met Major Miller and he had been searching for a group of children just as he was now. He blinked away the memories and tried to focus on the present.

The tentacle moved again above his face and its probing tip touched his right cheek. The flesh of the tentacle ruptured, pouring forth green ichor and the entire jointed arm convulsed. The eye stalk quivered and gobs of flesh broke loose from the monster, plopping onto the table. The flesh sizzled where it touched Steel's skin. The nightflies hissed and jerked back from the table.

Above him, the probing monster gyrated and pulsed in its shadows and a thousand inhuman mouths wailed in pain and terror. It receded into the dark shadows and the stench of its decaying, dying flesh engulfed Steel. He gagged and bile filled the back of his throat.

"Interesting." The silhouette of a man appeared at the foot of his table. "What have we here?"

The nightflies flocked to him like grubs to vomit. "My pretties, relax. I will not allow harm to come to you!" He glanced up into the dark shadows of the ceiling. "It will survive. I have given it a remarkable genetic ability to regenerate any form of human tissue." His face remained hidden in silhouette. "The eyes came from the DNA of Ralston Mead. And, there have been dozens of others who contributed their genetic material so that my Probenosticon can live. When it has recovered, I will find out how you managed to hurt it. Quite generally, it is the other way around. No one has ever withstood the ministrations of my creation much less injured it. Who are you?"

Steel tried to speak but the paralytic force kept him splayed on the table. The man raised a hand and touched his right ear. "Excuse me. I have to take this."

His face was hidden in darkness as he nodded. "Good news, Barnard! You are sure the satellite feed is accurate. She has found the artifact? That is most fortunate." He touched his ear again. "Well, Vivian, this paltry prize I was searching for is no longer of any importance. An artifact for which I have been searching has just been located and we have a new destination."

Steel's heart constricted in fury as Vivian Ketrick stepped out of the nightflies. She glanced at him. "A new destination?"

"Reyjacklic."

One of the nightflies stepped forward. Its head did not spin and resembled the stunted, deformed head of a human. "Reyjacklic was one of my first experiments. I cloned Ralston Mead's head onto its body. It did not take, unfortunately and poor Reyjacklic is imperfect. He will make sure that Vivian takes both of you to the cargo hold."

"And if they attempt to escape?" She kept her gaze locked on Steel.

The man pulled a pistol from his jumpsuit. "Use this. It is an inertia gun producing a stunning effect. And, it will not work on me, Vivian. Don't get any ideas. When we arrive at our final destination, so will they." He gestured to Steel and Knight. The man disappeared behind her along with his nightflies.

Reyjacklic hovered near the table. Its wings did not

vibrate as cleanly as the rest. He had only three arms and one of them was stunted and deformed. "I shall obey my master." It hissed.

"Then, you will obey me also?" Vivian asked.

"Only if such comports with the commands of my master." It said.

"Get them some jumpsuits."

"Why?"

"They shouldn't die naked. Give them some dignity. I'm sure you understand what it is like to have lost your dignity?" Vivian said.

Reyjacklic disappeared into the darkness and Vivian bent closer so only Steel could hear. "This one is bad, honey child. The worst so far. When we get to the cargo hold, take the gun from me. It will be your only chance at escape. Understand? I'm here to stop the tenth demon and I can't have you helping me this time. Got it, sweetheart?"

Chapter 14

The pale walls of the tent billowed around Dr. Cassandra Sebastian as she stared at herself in the mirror. "Of all the times to change my hair!" She grabbed an old fedora from her sister's desk and pressed it onto her head. "Now, they'll think I'm Renee."

"I'm not so sure about that, Senorita."

Sebastian spun around. A portly man wiped sand and sweat from his forehead. His black hair was wild and merged into the beard that covered his lower face. He wore a khaki shirt and jeans.

"Juan? I have to pull this off. I can't let Renee get her hands on it."

"It is her dig, Senorita."

"And, you are being paid handsomely to make sure I knew when she found it." Thunder echoed down the canyon walls around them. Sebastian brushed aside the tent flaps and studied the sky. Plum colored clouds billowed over the closest mountain range and churned toward them. "A storm is coming."

"Yes. A bad one. Fortunately for you it will keep the Federales busy on the north shore. The drug cartel boats come during the storm when the radar is spotty."

Sebastian nodded. "In the rain, the other workers will think I'm Renee. This can work to our advantage. When did you find it?"

Juan rubbed his beard. "Last week, we found the ruins from the Jesuit mission. Your sister was right. The Jesuits hid their mission in this valley while they helped the Seri Indians learn agriculture. Does this look like an island for growing things? No wonder they all died out."

"And, no one is living here now?"

"No, Senorita. Hikers and campers have to get a permit. I made sure my friend in the government blocked those permits for a month. By the way, that will cost you more."

"I don't care. How was Renee getting it off the island?"

"A ship is waiting at the only dock. We were to put the artifact in a truck and take it overland to the ship. The pilot works for me, Senorita. All I need to know is where to take it."

"I'll give the coordinates to your pilot. I have my own ship on the way from Acapulco. But, we have to hurry. Renee is coming from Florida and she will be here within a few hours. I was fortunate to be closer by." Sebastian jerked as a barrage of rain pummeled the tent. "Where are we on excavating?"

"The object was at the bottom of a fifty foot well carved into the rock. The Jesuits must have taken years to transform the well. They carved stairs into the rock wall. At the bottom, there is a shrine of sorts with huge metal doors. That is where we found it."

"Let's go." Sebastian hurried out into the storm. Rain stung her face and the wind whipped through the canyon toppling tents and swirling debris in its wake. The gale snatched the hat from her head. She followed Juan across the rocky canyon floor to a broken stone archway. The path led through the crumbling remains of an adobe village to the center of the mission. A huge metal tripod sat astride the yawning hole into the earth. Suspended from the apex of the tripod hung a huge wooden square crate over fifteen feet in width and three feet in thickness.

"You have already crated it? I was hoping to see it." Sebastian shouted.

"If you want this crate on the ship before Dr. Miller arrives, we must move quickly." Juan screamed into the rain filled wind. Clouds churned above them and lightning spiked into the nearby mountain ridge. Thunder echoed down the canyon. The crate swung from chains and ropes like a platform suspended from the tripod.

"They have to be more careful!"

"Senorita, the disc is made of metal. It will not break."

"Metal?" Sebastian stumbled in the wind. Juan grabbed her arm and pulled her back from the pit's edge.

"Some kind of shiny metal."

"Most likely it is polished platinum." Someone said behind her. Sebastian whirled. Anthony Cobalt stood against the wind. He smiled.

"I once visited a dig in the Negev desert. A Palestinian archeologist claimed to have found one of the early manuscripts of his religion that were thought to be destroyed by the fathers of his faith. His naiveté proved to be his downfall. One of his workers detonated an explosive device in a suicidal gesture. I was fortunate to be up the hill in the helicopter about to land. The explosion killed everyone in a one-hundred-foot radius and collapsed the cave in which the documents were hidden. I barely escaped with my life." Cobalt moved next to her and glanced down into the pit.

"What are you doing here?" Sebastian shouted above the storm.

"Well, I am not planning on committing suicide. But, my dear Cassie, you surely didn't expect me to ignore your satellite findings. Especially when we are both looking for the same thing."

Cobalt nonchalantly shoved Juan over the edge of the pit and his scream was lost in the wind. Cobalt grabbed at her and Sebastian dodged his hands and jumped through the rain filled air. She landed on the edge of the crate. It swayed beneath her weight and she slid off the side. She snared one of the dangling chains and hung precariously above the open pit.

The disc shaped craft came out of the angry clouds. Its undersurface gleamed with green light and lightning played across its surface. An iris opened in the center of the underside of the disc and it centered itself over the apex of the tripod. Two men tumbled out of the iris.

The nightfly fumbled with Steel's restraints. The nightmarish face gazed at him with cow eyes.

"Reyjacklic, you are an outcast from your own kind?" Vivian said.

"Yes." It hissed.

"Will you pledge yourself to me? Will you serve my demons?" Vivian reached out and touched one of the thing's arms. Its skin was cold, mucoid. She shuddered. Reyjacklic stiffened and its eyes widened.

"I sense you are in league with a Vitreomancer demon?"

"Yes." Vivian fought back nausea at the stench of the thing's breath. "He is my head demon."

"Then he is very powerful. Very powerful."

"Powerful enough to reverse your deformities?" Vivian asked.

"Possibly."

"Then pledge yourself to do my bidding, Reyjacklic. And, when this is over and Cobalt has discarded you, I will make you whole."

"I pledge myself to you." Reyjacklic said.

"Vivian, what is going on?" Steel finally found his voice.

"The tenth demon, love. Do you even have to ask? I'm here to stop him and I have very specific orders from the Council not to involve you." She threw aside the last restraint. Steel bolted up from the table and grabbed her arm and pulled her against his bare chest. She winced at the heat of his skin. She inhaled the fragrance of his sweat. She blinked and tried to clear her thoughts.

"Why is that every time I turn around, you are involved in my life?"

Vivian jerked her arm out of his grasp. "Honey child, it wasn't my idea. You seem to have a friend on the Council. I'd just as soon as let you get tossed out of the ship." She looked away from him and tried to erase the image of his turquoise eyes from her mind. What was happening? She hated this man. Detested him. She stoked the hate and fury and ripped away at Knight's restraints.

Steel leaned over Knight. "Arthur? Can you hear me?"

He groaned and his eyes rolled. Vivian motioned to the nightfly. He handed her a jumpsuit. "Here, get him into this."

Steel helped Knight sit up and slid his legs into the one-piece jumpsuit. "Come on, Arthur, stand up and get this on. Where are we headed?"

"We're going somewhere to pick up an artifact for Anthony Cobalt."

"Cobalt?" Knight whispered. "Not him. No!"

"Yes, he's after some kind of artifact. I've seen the reconnaissance already. Some kind of metal tripod over a pit. We're going to the cargo hold and there will be this huge opening." Vivian took the inertia gun from the nightfly.

"Steel, take this from me and stun me then jump for the tripod. It's your only chance. Once you get out of the ship, you are on your own."

"I could take it from you now." Steel growled as he helped Knight into the jumpsuit. The nightfly stepped between Vivian and Steel.

"I'm afraid this one has pledged himself to me. It seems he is a deformed creature cast out by the rest. I like outcasts, Steel. They become my friends."

"Like Armando and Bile."

Vivian tensed at the name. "Bile, yes. Anyway, we are on some island in the Gulf of California called Isla Tiburon. Cobalt said we were heading for an old mission hidden in the mountains. So, who is your friend?"

Knight straightened and moaned. "Arthur Knight. What's left of me."

Vivian laughed. "The kid's father? Honey, you are supposed to be dead."

Knight glared at her. "Exactly what I wish for you. If I had the strength, I'd kill you where you stand for what you've done to Josh."

The nightfly stepped toward him. Vivian motioned

toward a door. "My friend will make sure you don't. Let's go." She disappeared into the shadows.

Steel shifted his weight to better support Knight and the nightfly tensed. The gun it had taken from Vivian looked like a twisted black vine with a glowing red tip. "I am known as Reyjacklic. If you try to escape, I will stop you with the inertia gun." The air shimmered between the creature's gun and Steel felt like he slogged through molasses.

"I get the point." He put Knight's arm around his shoulder. "You good to go?"

"I'm getting my strength back." He said. "I'm not that weak." He whispered to Steel.

If what Vivian had said were true, Steel would have to get the gun away from the alien before they could escape. And, Knight would have to hold his own. The nightfly motioned ahead into the shadows and they approached a ramp leading downward. They emerged in a corridor opening into a large cargo hold. Rain and wind blasted them and Steel looked through the open bottom of the ship at a vortex whirling around a deep pit in the ground. A crane towering above the pit reached to the level of the ship. Dangling in the air a wooden crate was ascending toward the craft in the center of a column of purple light.

Steel glanced to the far side of the opening and spied Vivian. She was struggling with the wind and her eyes met his. She nodded toward the opening. They were supposed to jump through this hole? Vivian hadn't given him a jumpsuit and he dreaded the damage that sand would do to his bare skin.

"I'll get the gun." Knight whispered.

Steel squinted against the blowing sand and saw a man and a woman standing on the edge of the pit. Knight rolled off Steel's shoulder and in one fluid motion, slapped the device out of the nightfly's hand and it skittered across the floor of the cargo bay. Steel scooped it up and pointed to the crate.

"Jump!" He screamed. Both men fell into the storm and landed on the surface of the crate. The crate teetered under their weight and Steel spied a hand gripping the edge.

"There! Someone is falling off the crate!"

Knight thumbed the button on the dampening field and pointed it up at the chain. The swaying motion slowed and Steel crawled slowly toward the edge. He grabbed the hand and pulled with all of his might. A face appeared over the edge. It was Renee Miller!

"We've got a problem." Knight shouted over the wind.

Steel glanced up. Standing at the edge of the iris was the man who had just been at the edge of the pit. "That's the man who tortured us."

The man gestured toward Knight and the dampening field device flew out of his hand and disappeared into the storm. The crate began to sway and spin.

The woman pointed to a stabilizing rope leading from the corner of the crate into the pit. "We have to get into the pit."

Steel snared the rope and handed it to her. He motioned to Knight and followed her down the rope into chaos.

Chapter 15

The woman from the crate ran across the floor of the pit and stooped over a body. Steel helped Knight down and glanced up at the huge ship hovering in the midst of the storm.

"That is Anthony Cobalt and if he is looking for the children, Steel, we are in terrible danger." Knight gasped for breath.

Cobalt stood at the edge of the open iris and gestured toward the cargo bay behind him. A huge, hulking mass of pustule covered flesh rippled over the edge with its appendages and folds marred by blisters and weeping sores. It fell, gyrating through the harsh purple light and fell into the pit. Steel grabbed Knight by the shoulders and pulled him away from the mass as it quivered and heaved, stalked eyes whipping through the air searching for safety. A mass of disjointed, uneven stalks shot from the bottom, lifted the huge maroon mass and skittered across the floor of the pit into an open doorway leading into an underground chamber.

"It doesn't like the light, I am afraid." Cobalt shouted. "One last gift." He reached into his pants pockets with both hands and took out handfuls of glowing, yellow beads. He tossed them through the air and they rained down upon the open floor of the pit. "When I fire up my ship, its energies will release the power stored in the Sunstones." He shouted through the wind. "I give you a choice. Remain in the pit and be cremated. Or, join my pet in the darkness of the underground chamber. Adieu."

The cargo opening shut and a pale, greenish light fell into the pit from the periphery of the ship. The undersurface began to glow with lavender light. Steel shoved Knight toward the chamber. "Everyone, inside now!"

The two remaining workers were first into the doorway followed by Knight, and finally Steel. He glanced back and the woman was cursing at the departing ship. He rushed across the uneven floor of the pit as the lavender light cascaded down toward them. He grabbed her around the waist and pulled her across the sand into the doorway.

"Let me go! He killed Juan!" She tore at his hands.

"Doors!" Steel screamed. "Knight, get the doors closed!" The pit filled with lavender light and the Sunstones began to glow. They hovered up from the floor of the pit. "What are those things?"

The woman pulled out of his grip and helped Knight and the other two men with the huge metal doors. They swung from the inside and Steel joined them.

"Sunstones." The woman said. "Cobalt's little invention. Crystallized power. There's enough of them out there to incinerated a city block."

The Sunstones pulsed with power under the onslaught of the lavender light. Groaning and sweating, they finally pushed the two doors closed. A huge, metal cross beam fell down from the wall.

"Secure the door." The woman screamed as a bright, yellow light began to leak around the edges of the door and through the crack in the middle. Steel grabbed the cross beam and put his weight beneath it. It slammed down into slots on the inner side of the door.

"Get back!" She screamed.

Light as bright as the sun burst through the cracks around the door and through the center seam. The air boiled with heat. Steel motioned them down away from the entryway to the lower level of the chamber. The group huddled at the base of the stone balustrade above them. Heat and light filled the chamber.

"Oh my!" The woman grasped.

The walls of the chamber were covered with murals and paintings. The light became so bright, they had to cover their eyes. Somewhere beyond them in the dark shadows of the chambers hallways they heard the creature roar with inhuman pain.

Abruptly, the light faded. Steel glanced over the edge of the stone balustrade. The metal door glowed red hot and the cross beam had melted, fusing the doors into one seamless lump of twisted metal.

"Well, good news is we are alive. Bad news is the doors are melted shut."

The chamber fell into darkness tinged by the slowly fading red glow of the doors. One of the workers ran up the stairs and shoved against the crossbeam.

"Stop!" Steel shouted but too late. The man screamed as he touched the cross beam. His clothes burst into flame. He ran screaming toward them and tumbled over the balustrade. Steel tried to stop him but he ran headlong into the back of the chamber. Like a living torch, his passage illuminated the murals on the wall and he bounced against the back wall of the chamber. The creature appeared from a fold of darkness and engulfed the man in its blob like mass. His screams were snuffed out just like the flame.

"Juan is dead. And now, Pedro!" The woman gasped putting a hand to her mouth. "Jesse, are you okay?"

"Si, Senorita. But, I am scared." The remaining man said from the shadows.

"Stay in the light and away from that thing."

"Diablo boca!" He whispered.

"Yes, the mouth of the devil." She wiped soot from her face and smiled at Steel. "You saved my life. Thank you. I remember you from Renee's office."

Steel studied her face. "You're Renee's sister!"

"Yeah, I beat her here."

"To take that artifact from her?" Steel asked. "What is wrong with you two?"

"Don't go there. I just lost the most important artifact of my life." Dr. Sebastian said.

The creature roared in the depths of the chamber. Darkness crept in on them as the metal door faded completely to ashy gray. "Does anyone have a light? When it gets dark, it'll pick us off one by one." Steel said.

Sebastian pulled something off of her hip. "My sat phone. I forgot I had it." She powered it up and a faint blue light filled the chamber. For a second, Steel saw the workman's empty eyes stare back at him from the man's blistered head, now imbedded on a bone like spike at the top of the creature's body. The creature scuttled back into the shadows.

"Can we call?" Steel said.

"We're 50 feet below the surface. No signal."

"Maybe I can fix that." Knight said as he squatted over a bundle of wires he held in his hand.

Sebastian glared at him. "Who are you?"

"I think you know."

The sat phone tumbled out of her grasp and her hands went to her face. "Arthur? Knight? Impossible."

"It's me."

She ran across the chamber and stopped just short of the man. She swung her hand around in a roundhouse and slapped him across the cheek. "You're dead! You're supposed to be dead!"

Knight rubbed his cheek. "I faked my death."

"But, Claire! And, Josh! How could you?"

"I had to protect them from Miller and Cobalt. You understand! It was the only way."

"Where have you been? All this time I've been looking for the throne and you were there watching over the children?"

"Is that all you care about? Your precious artifacts?"

"Yes, and you know, don't you? You know where the Bloodstone is."

Steel picked up the sat phone and stepped between them. "Stop it! We don't have time for this. Renee left this afternoon and I assume she was headed here. Contact her with the sat phone."

Knight took the sat phone. "This chamber is airtight and we won't last but a few hours. Particularly with that blob thing trying to pick us off. These wires lead to the surface and I think, if they are intact, I can use them to extend the antenna."

He sat down and began working on the sat phone. It's meager light cast moving shadows across the wall.

"I need a moment." Sebastian walked away from them and sat down against the stone wall.

"Stay in the light." Steel said. He squatted beside Knight. "You know Dr. Sebastian?"

"Yeah, I worked for Renee, remember. I've been protecting the children and there are many times I had to steer Cassie astray."

"Cassie? You're on a first name basis?"

Knight looked up from the wire. "Look, we've got a lot of history. I don't have time to deal with that right now. We have to get out of here and stop Cobalt. If he gets his hands on the children, heaven help us." A beep came from the phone as he twisted the last wire onto the back of the sat phone. "Got a signal!"

Sebastian bolted up from the floor and took the sat phone from Knight. She dialed a number and spoke over the phone. Steel watched her pace and mumble into the phone. She was resilient, he had to give her that.

"Renee, listen to me. Stop shouting! He took it! Yes, and we're stuck at the bottom of the pit. When will you be here?" She paused and looked up at Steel. "The Federales? Bribe them! Give them lots and lots of your precious money! Isn't that how you got out here in the first place?" She gasped and looked at the phone. "She hung up on me. Can you believe that?"

"Really? How long before she gets here?" Steel asked.

"An hour, maybe two at the most. Seems the Federales saw the fire and are coming to investigate."

"May I?" Steel motioned to the sat phone. "I need to call Cephas."

Josh stumbled and landed on the rough rock. Pain lanced into his kneecaps and he fought back nausea. He gulped for breath and looked up at the man standing in front of him. He was tall and backlit by the distant lights in the cavern. His head was haloed against the harsh light and he raised a cane into view. A wolf's head stared back at Josh from the cane.

"My young friend, it is time for you to feed my sheep." The man said in a heavy Eastern European accent.

Josh shook his head. This couldn't be happening. He had escaped from Rudolph Wulf and his vampires. Someone jerked him to his feet and pain shot through his right shoulder. He was shoved forward down the rocky path toward the distant platform above the chasm that led deep into the bowels of the earth. The silver spears glinted in the light and the cage that would hold him dangled above the spears.

"I won't do it!" Josh pulled himself out of the grip of the person behind him. He whirled. It was Jonathan Steel. "Jonathan?"

"I'm growing tired of you, Josh. It is time for you to go."

Josh glared at the man's turquoise eyes. "Fine! Bro, I knew you hated me! If only my father and mother were here."

Someone stepped into view from behind Steel. His face was hidden in shadow and his eyes caught the reflected light from the cavern. He recognized the eyes and his heart raced.

"Dad?"

The man turned his back and started walking away. Josh reached for him and Steel blocked his way. Rudolph Wulf's iron grip tightened on his shoulder. A bright, burning light filled the darkness and he watched his father walk into the sun. His skin blackened and his hair burned away and the flesh became ash leaving only his father's skeleton. The dark silhouette of his bones disappeared into the blinding heart of the sun. He screamed and screamed as the heat and light engulfed him.

Josh sat up in the car seat, his body soaked with sweat. He glanced at Cephas Lawrence behind the steering wheel.

"Bad dream? I told you we shouldn't have had tacos." Cephas kept his eyes on the road. He had both hands gripped to the top of the steering wheel and he was so short, his head barely cleared the top of the wheel. Any other time and Josh would have found it funny. But, the vestiges of the dream still filled him with dread. In the back seat of the SUV, Theo snored.

Josh rubbed sleep out of his eyes and pulled the sun guard down to block the hot rays of the setting sun. They were traveling west and the heat of the sun had exacerbated his sunburn.

"It was a bad dream about Jonathan and my Dad. Dad is gone. Mom is gone. All I've got left is you and Jonathan."

He watched something move across Cephas' face. The man blinked and pulled the car over off the interstate and up to a rest stop. He turned off the car and sat there quietly as the heat of the late afternoon quickly soaked into the interior.

"Josh, you must accept the fact that Jonathan has your best interest at heart. He would never do anything to hurt you."

Josh sighed and sat back. "I know, Uncle Cephas. I just wonder what would happen if my father were here. I miss him. He liked to play ball in the back yard. Sorry, Uncle C, but neither you nor Jonathan play pitch. And, it's been so long since Dad died. Sometimes, I can't even remember what he looks like."

Cephas patted his arm. "Things would have been much different if your father had survived. But, you can't live your life based on 'what if', Josh. You're here, now and you have to make decisions on what happens today. Living in the past can't help."

"What's up, Papaw?" Theo leaned forward from the back seat. "Need me to drive?"

"That would be most welcome although I doubt very seriously if my snoring will keep you awake. Your snoring would keep the dead awake."

Josh opened the door. "Dude, I've got to pee." He walked across the dried grass of the rest stop and entered the men's restroom. It was dank and stinky but deserted. He finished his business and stepped up to the sink. There was a warped piece of shiny metal instead of a mirror hanging above the sink. Josh rinsed his hands and reached for the nearby hand dryer. Something moved in the reflection of the room behind him. He glanced up at the metal mirror. A woman dressed in a white dress with long, flowing hair stood behind him. He whirled to face her.

"Don't be afraid, Josh." The woman said.

"What are you doing in here? This is a men's restroom." His heart was racing and he leaned back against the sink.

She smiled. "I'm just here to tell you to trust Jonathan Steel. More pain and suffering await you and Cephas but you must be faithful. Remember, faith and fear cannot co-exist."

"How do you know who I am?"

She pointed over his shoulder and he turned around to look back in the warped metal. The room was now empty. He whirled again and the woman was gone. Who had she been? For that matter, was she even real? Maybe he was still half asleep. He turned back to the sink and poured cold water over his head. He washed his face and slowly looked up at the mirror. The room was still empty.

Josh ran out of the restroom. Cephas was walking around the vehicle and Theo slid behind the wheel. Cephas stopped.

"Josh, you look like you've seen a ghost."

"Maybe I did, Bro."

Cephas' cell phone rang on the front seat and Josh leaned in and snared it.

Steel dialed the number into the sat phone and waited.

"Hello." Josh said.

"Josh, put Cephas on the line."

"Dude! Where are you? We've been driving all morning back to Shreveport and I had just met these hot girls on the beach." He paused and Steel could hear the roar of the passing highway. "Are you okay?"

"I'm fine. Let me talk to Cephas."

"Yeah, right!"

"Josh, put Cephas on the phone now."

"Uncle Cephas, it's Jonathan."

The line rattled and Steel heard Cephas' heavy breathing. "Where are you?"

"Long story, Cephas. You were right about the tenth demon. He has the children and he abducted me in a space ship, a kind of flying saucer."

"Where are you?" Cephas whispered.

"Tiburon Island somewhere in the Gulf of California at Renee's dig."

Cephas gasped. "The artifact?"

"Anthony Cobalt has it now. Cephas, what is happening?"

Cephas was quiet for a moment. "He will need the other two pieces of the artifact. Jonathan, we must stop Cobalt from acquiring all of the pieces."

"What will happen, Cephas?"

"That I do not know. But, I do know that such an object once assembled will give him great power."

Steel nodded. "Okay, let me talk to Theo."

"Chief, you okay? I ought to kick your butt for going off without me." Theo growled.

"I didn't know I was going into a firefight, Theo."

"My point exactly."

"Listen, Cephas can fill you in but I need you to keep an eye on Josh."

"It's the next one, isn't it? Numero ten."

"Yeah, Theo."

"Anything I need to look out for?"

"Little green men and flying saucers."

"Any other white dude tell me that and I think he nuts. But, coming from you, it don't surprise me. Watch your six, man."

Steel ended the call. The sat phone light wavered and it beeped. Sebastian studied the readout.

"Hooking it to that wire must have drained the battery. We only have a few minutes of power."

Something rustled deep in the chamber. The pale blue light began to fade and Sebastian's face with it. "Jesse, do you have any kind of weapon?"

"No, senorita. My pen knife will not stand up to that devil."

"Any other source of light in here?" Steel asked.

"We have some flashlights and flares in the back of the chamber. They are in a case against the back wall, senor."

"So is that thing." Knight's face appeared in the shrinking circle of light. His eyes were dark. "I don't think I can face it again."

Steel took the sat phone from Sebastian. "Then, I need to take that thing out."

"You saw what that thing did to me, Steel. You know what it is capable of." Knight said.

Steel nodded. "I also saw what happened when it touched me. I don't think it likes me." He jumped over the stone balustrade and into the darkness.

The meager light from the sat phone was useless by the time he reached the back of the chamber. The shadows shifted and he smelled the thing as he drew nearer. From the inky blackness a glimmer appeared floating in the air and the head of the workman hovered on a bony stalk. The man's face was distorted by tissue from the thing and a third eye had sprouted from the forehead. The man's dead eyes were white and milky but the third eye was huge and bulbous. It rotated as it studied Steel. The rubbery tissue had covered the man's lips and now contracted and moved. It spoke.

"What will you do with me, human?" Its guttural voice grated on Steel's nerves.

"Destroy you." Steel said.

"You do not have the power." It hissed.

"I think I do." He tossed the sat phone aside and grabbed the man's head with both hands and pulled. Where the thing's tissue had grown, flesh bubbled and blisters formed, filled with bloody goo and pus. They burst and the fluid ran over Steel's hands. The bony stalk disengaged and Steel fell back with the man's head in his hands. He hurled it aside and dove for the creature. Tentacles shot out of the darkness and wrapped around his neck and his waist. They began to smoke and bubble even as they contracted. Steel's consciousness wavered. If the creature killed him before he killed it what good had he accomplished?

Knight appeared beside him with a huge stone held above his head. He hurled it onto the tentacles and they tore away. Steel rubbed his neck and glanced at Knight. "Thanks. I've got it, now."

Steel dove into the darkness; into the very mass of the thing and was instantly engulfed in the fetid odor and rotting flesh. A thousand mouths formed and screamed in unholy terror. Steel tore and ripped at the thing's flesh; breaking bones and joints; pulling asunder gaping jaws and twisting away eyeballs. The thing tried to smother him; engulf him but like some dying, drying ameba it lurched and quivered and smoked and disintegrated under the harsh light of Steel's soul. It died at the hands of a man of God. It settled and pooled around Steel's knees a useless, rotting pool of goo. Finally, the goo hardened and flaked and fell into ash.

Steel blew the rancid ash from his mouth and nostrils. Greasy soot covered his bare chest and smoke hung thick in the air mixed with the odor of burned flesh.

"We need to find the lights." He spoke into total and complete darkness.

"I know where they are, senor. Is that devil dead?" Jesse said from a distance.

"Nothing but ashes, Jesse."

Jesse ran along the far wall and collided with something metal. "Ah, here's the equipment chest."

Light flooded the chamber and Steel squinted. Jesse carried two bright lights on a stand to the center of the chamber. "The batteries will last about an hour, senor."

Cassie appeared behind him and her face turned up to the walls. Steel brushed the remnants of the foul creature from his chest and turned around. The walls were covered from floor to ceiling with faded, color murals. An image of a huge throne covered the entire back wall. It was shaped like a sitting angel with two wings unfurled upward behind him and a second set of wings wrapped around its feet. But, this angel had two sets of arms. One set stretched above his shoulders holding a gleaming, crystalline sword pointed toward the heavens. The second set of arms reclined on the legs providing the arm rests for the throne. But, the most peculiar thing about the angelic throne was the lack of a head. In place of the head, a mask floated above the neck. Elliptical eyes of blackness stared into space above a thin slit for the mouth.

"Oh, my! The Fallen Throne!" Cassie exclaimed. She started toward the image and stopped at the edge of an open depression in the center of the chamber. The depression was perfectly circular and about fifteen feet across and a foot deep.

"Jessie, is this where the artifact came from?"

"Yes, Senorita. We put it in the crate."

Steel put a hand on Sebastian's shoulder. "What was here?"

"The Infinity Disc. And that throne, if it were here, would have sat on the far edge of the disc."

"Only the disc, Senorita." Jesse said.

"The angel has no head." Knight said. "Is it true, then?"

"Is what true?" Steel asked.

"The head of the angel is the Bloodstone." Sebastian said.

Chapter 16

Vivian watched the earth rush beneath them on the monitors in the control room. The nightflies, as Steel had called them, stood before consoles and in the center of the chamber, the glass tube contained the hideous spiral of the pilot of the ship. She was unsure of its pure substance, but the evil it emanated disturbed her. At her urging, her demons had grown quiet and subdued in the presence of the creature. The fluid in the cylinder was milky and occasionally a ragged strip of flesh would caress the inner walls of the cylinder. When it did, the nightflies froze at their consoles. Even their rotating heads stilled. Then, the flesh would disappear into the milky madness and they would resume their movements. Cobalt appeared behind her resplendent in linen pants, a flowered shirt, and a loose linen blazer.

"What is in the cylinder?" Her inner demons grew more agitated.

"The pilot, so to speak. An unspeakable abomination in this universe, Vivian. Ah, if I had time the tales I could share with you. Suffice it to say, it came from before time and space, from when and where we were created before the Creator designed this awful universe for you puny humans." Cobalt smiled and arched his hairless eyebrows. "It is very angry. And, that anger powers my propulsion system."

"Supernatural power."

"Of course. You don't think anything in this universe could allow us to travel as we do, do you? Our maneuvers defy the laws of physics. But, they do not defy the supernatural!"

Vivian grimaced and rubbed her head. Keeping her demons subdued taxed her strength.

"Headache, my dear? Keeping your demons still against the pure evil of that thing will result in futility. Why don't we return to the lounge?"

Vivian nodded and followed him down the corridor. "Where are we headed?"

"Back to Area 613. I must rendezvous with the other ship and pick up the children." He went behind a wet bar and poured amber fluid into a snifter. "Brandy?"

"Please." Vivian relaxed and the demons wallowed in her mind. Their sharp edged insanity almost engulfed her until Vitreo chastised them into submission. She thanked it and then cursed it for being so weak. An image of its pallid, white orbs floated before her eyes as it recoiled. As the eyes faded, it laughed.

Cobalt handed her the brandy. "The children will become my new army, Vivian. Their powers are not of this earth and soon they will serve me."

Vivian sipped the brandy. "Children? What children?"

"In due time. They will worship me, Vivian. Wait and see."

Vivian watched the clouds whip past the ship. "You enjoy being worshipped, don't you?"

"Once I acquire the other two pieces of my artifact I will sit on the Fallen Throne and all will worship me, Vivian. It will not be the first time. Would you like to hear the tale?"

Vivian glanced down at the ring on her hand. The crimson stone was dark. Perhaps this was a story not yet chronicled by the Grimvox.

"By all means, amuse me."

"I have seen many hosts over the millennia." His voice changed and his lavender eyes glowed with supernatural light. "One such host was a woman of great beauty and power."

Semikabal let the wind blow through her hair as she

surveyed the many flickering lights of the kingdom below her. It had taken three lifetimes to build the great tower of Babel and her host was the fourth in that period. Only she knew the true meaning of the name, "gate of the god". Her husband, King Nimrod was so focused on his archery and his prowess as a hunter, he scarcely understood why Semikabal had urged him to finish the tower his ancestors had begun.

"You appealed to his ego."

Semikabal turned her back on her kingdom and squinted into the darkness at the top of the tower. "Who is there?"

A man stepped out of the shadows dressed in a long, flowing green robe of the priesthood. His dark hair writhed in the wind and his eyes burned. "It is your nemesis, Semikabal."

She swore and pushed her unruly hair out of her face. "Donnota? Why are you here? I thought I had banished you from the kingdom."

Donnota smiled and moved to the edge of the tower. "There is no place outside the kingdom for me to go. You have sufficiently blinded the king to the commands of God to subdue the Earth, to move out from here and occupy all of the lands He has created. There is but one place man can live. In his kingdom."

Semikabal moved up beside the priest. "But, if we send our people out into the world, then we will have no subjects. Nimrod is a rude hunter but he is ambitious as all men are. I appealed to his sense of ego and arrogance. As I do, he desires a kingdom, a world to subdue."

"Thus, this tower?" Donnota looked at her. "What is the purpose?"

"It is a gate of the god of the sky. It symbolizes Nimrod's ascension to godhood."

"Then, he is the first idol?" Donnota asked.

"The first of many. You can't expect man to continue to worship a God who kills all of humanity but a handful in a great flood. What a cruel God that is."

"If not for you and your kind, the flood would not have been necessary. You corrupt mankind. You appeal to their

rebellious nature and turn them against God. You make them look inward." Donnato said.

"Ah, but only to make them look outward for something to believe in besides their invisible God." Semikabal smiled.

"He was not invisible in the garden. Man chose to sever the sanctity of that relationship. He chose knowledge over obedience."

"And, now I will insure that man looks to the heavens beyond God for meaning and sustenance."

"Why deceive them so?" Donnato brushed his dark hair out of his eyes. "Ah! I see! Are you still clinging to that hope that somewhere out there your kingdom awaits you?" Donnato motioned to the stars gleaming above him.

"It is there! I tell you that somewhere in the heavens a world awaits my coming."

"And yet, your master sends you on minor missions of confusion and deception." Donnato's green robes rustled in the night wind. "I think I understand what you are doing, Semikabal. You are playing a very dangerous game with your master. You urge the building of this tower and you infer to your master that its purpose is to introduce the very concept of idolatry. You have convinced Lucifer that this tower represents man's attempts to replace God with an exalted human, Nimrod." He stepped closer to her. "But, the real reason is to reach the sky. To point man to the stars and to a hope in some magical world that awaits them."

"Yes!" Semikabal's throat warmed with excitement. "And, one day, when mankind leaves this world behind, I will be waiting for them as their new god. Once they look upward and begin to study the night sky and see the unreachable stars, I will turn that desire to my advantage. Once I have my world, I will be their new god and they will come to me and into my welcome arms."

"Yes, man has always looked to the sky. They think it is the home of God's throne."

"Before the flood, their knowledge was purer and more closely applied to the knowledge they learned from God. They had the capacity to build vessels to reach the stars, but in their

stubbornness and arrogance, they have lost their long life spans and have lost their knowledge. I must help them to recapture it." Semikabal pulled her thin robe about her. "This tower is the first step."

"And, if Lucifer learns of your deception?"

"It will be too late. He is as arrogant as man. I know how to appeal to his ego."

The sun had risen by the time they ended their conversation. Semikabal looked down upon the thousands and thousands of people far below who had gathered for the dedication of the Tower. "See, already they gather like sheep at my feet. They are willing to do my bidding."

Footsteps sounded behind her and she turned to see her king appear at the top of the tower. He was surrounded by his entourage of servants and advisors. He was tall and muscular with a trim, black beard that skirted the edge of his jaw. His hair was braided and fell down his back. He wore a leather skirt and the bare muscles of his chest bulged with power. There at his waist was the crystalline sword. He refused to go anywhere without the vile thing! His dark eyes took in Donnato and then traveled over Semikabal. Naked lust stirred within him. Her power over him was complete and there was nothing Donnota could do.

"My queen, have you been here all night?" Nimrod took her by the waist and pulled her to him. She inhaled his masculine scent, the musky nature of his sweat and faint odor of fish on his breath.

"Yes, my king. I have been waiting for you to come and assert your power. It is time to become more than just a man."

Nimrod raised an eyebrow and smiled. "And, I have you to thank, my queen. My bow!" He turned to his entourage. A servant ran forward with Nimrod's mighty bow. Nimrod lifted it into sight, the rising sun glittering off of the golden highlights on the wood. The string was gilded with gold threads. He pulled the string back and stretched out his arm. The muscles rippled in his chest and arms. He smiled at Semikabal and stepped forward into view of all the people far below him.

Donnato stepped around Semikabal to catch Nimrod's

attention. "My king, I advise that you not pursue this course of action. God has blessed your reign and it would not be wise to turn against him."

Nimrod glanced at Donnota and sighed. "I am weary of your pleadings, Donnota. My queen has shown me the path of wisdom. Am I not the greatest hunter in the world? Should I not only be obeyed but worshipped?"

Donnota glanced at Semikabal and backed away. She smiled. She had won. Nimrod had made the Choice.

"My people, I come today to proclaim myself not only your king but your god." His powerful voice echoed down the tower to the masses waiting below. "I have built this tower to the stars to proclaim my mastery of the earth and the heavens. God has not chosen to stop me. Today, I will demonstrate to him who is the mightier of the two."

He snapped his fingers and the servant ran forward. He took the golden arrow from the servant and held it up. It caught the morning sun and glittered with power and light. "This arrow will pierce the heart of God and I will assume the throne of the heavens." He shouted. The people below went wild, their voices traveling up the side of the tower chanting his name.

Nimrod motioned to Semikabal and she stood at his side. Her heart raced with eagerness and anticipation. The common rabble below would do whatever Nimrod told them. The last star to be conquered by the rising sun star glittered on the western horizon.

Nimrod notched the arrow and pulled it back. He pointed it upward toward the heavens. He released it and watched the golden arrow fly into the clear and cloudless sky. The bolt of lightning appeared out of nowhere and struck the flying arrow, bursting it into flames.

Nimrod gasped and the people below grew silent. Donnota stepped up beside him. "God is not happy with you, Nimrod. He has told me to tell you that all of your people will be confused."

Nimrod shook his head. "I do not understand."

Semikabal stepped between them. "Do not listen to him, my king. Get another arrow."

Nimrod looked at her with horror and began to speak nonsense. Nimrod shouted to his servants. They looked at him in confusion and a different language issued from each servant's mouth. Below, the people erupted into chaos and confusion. Their voices climbed the tower, confused and conflicted.

"What have you done?" Semikabal pushed past the confused Nimrod to Donnota.

"I have done nothing." Donnota smiled. "God has given each man a new language to speak. If man will not willingly go out and claim the world, then God will make them do so without a choice. Now, they will leave this area and subdue the planet. You see, when too many of them are gathered together in one place and united in one cause, they will always gravitate to evil. I'm afraid your plan to conquer the stars will have to wait. If they can't understand each other, they can't build another tower."

Semikabal clinched her fists and swore. She looked down upon the people and watched in horror as they turned on each other. Some pointed to the tower and others to her. They began to climb the tower, tearing it apart stone by stone. "No!" She screamed.

Donnota shrugged. "I tried to warn you. You might want to disappear before Lucifer shows up. He won't be happy."

Donnota disappeared from sight. And, as the people swarmed up the tower to destroy their king and his dream, Semikabal collapsed into a heap. The being who would become the tenth demon left her body and stood on the top of the tower. His skin was as black as onyx and his eyes gleamed with a hatred and fury that scorched the earth around him. His huge, crystalline wings unfurled and he launched himself into the morning sky even as the last star disappeared on the horizon.

Cobalt was silent and he sipped his brandy. An amber drop trickled down his chin and he wiped it away with a flick of his hand. "My first great defeat, Vivian."

"I'm surprised you admit to failure." Vivian said.

"Failures are necessary and teach us how to be victorious the next time." Cobalt said.

"What is in the crate?"

Cobalt swirled the brandy in his snifter. He studied the liquid with his lavender eyes. "Have I told you my host's predecessor worked with the KGB? In fact, he knew Rudolph Wulf. As a member of the KGB, he was a brilliant tactician and received his education in the United States. He was a child planted as part of a sleeper cell by the Communist regime."

Cobalt stood up and paced behind the couch. "His passion was evolution, Vivian. He became one of the leading geneticist in the country. His goal was to unravel the secret of the human genome. And, with help from his supernatural connections, he managed to develop one of the first gene splicers in history."

Cobalt leaned forward and his lavender eyes glowed with fervor. His cold, brandy tainted breath made her shiver. "Imagine, Vivian, the seduction of fame and fortune such a device provided. He could accomplish what would take scientists another twenty-five years to even begin." Cobalt pulled away. "But, his devotion was to the KGB and the Soviet empire. If they knew the device existed, they would take it and destroy him, his usefulness at an end. If he made the device known in order to enjoy the fame of his discovery, he would betray his native country. What would he do?"

Cobalt had walked around the couch and now stood between Vivian and the window. "Do you feel that tension, Vivian? Do you see the conundrum? Our base demonic nature is to foment chaos, to disseminate disorder. And yet, there are times when we see a glimpse of the creative nature of the Creator. We feel the tug, the pull, the seduction of, shall I say it, good. Just as my former self was conflicted, so am I. Vivian, I am at war with myself. How much do I tell you? Will you go running to the Council and tell them I have 'found the gene splicer', so to speak? Or, will you allow me to enjoy the fruits of my labor? You heard my story. Will you allow me victory or will you oppose me as my ancient antagonist has done?"

"I don't know, Anthony. I want a seat on the Council.

That is all I desire. I take it your predecessor never turned his invention over to the Soviets?"

"He, that would be, we, destroyed it. Our triumph was nullified in the face of the length of my plan that has stretched from the beginning of mankind into the future. If anything, Vivian, I have learned patience." He threw back the remainder of his brandy. "Now, what of this man you allowed to escape?"

Vivian gasped. "I don't know what you're talking about?"

"Lying to me is pointless. What is his name?"

"Jonathan Steel." Vivian swigged the rest of her brandy. "My nemesis."

"And, yet, you let him live along with his companion?"

"I was told to do so by the Council. They have plans for him."

"See, Vivian. You, too, are conflicted. You desire great power. Your demons grow ever more restless and hungry. Do you obey the Council? Or, do you obey your own desires?"

Vivian averted her gaze and studied the brandy snifter. Did he know about the thirteenth demon? Did he really understand how conflicted she was? "We are in dangerous territory." Vitreo whispered behind her eyes. She could feel him looking through her orbits with his ghoulish, white orbs. She closed her eyes.

"Silence! I have to be very careful here." She chastised Vitreo. His laughter echoed in her mind.

"Jonathan Steel." Cobalt said. "His soul is filled with light, Vivian. He is a special creature of the Creator, I fear. I've never met anyone like him. Well, maybe one!" He wiped a drop of brandy from his lips. "You see, my creature, my probenosticon could not touch him. How is it you can touch him?"

"What?"

Cobalt lifted a hairless eyebrow. "There are those who have sworn their allegiance to the Creator, Vivian. That allegiance is so profound, we cannot abide being in their presence. You host seven demons, Vivian. It is a task many have never mastered. And yet you not only aid this man of God," he grimaced at the mention of the name, "But, you touch him and you are not harmed."

"I don't know how. I hate him."

"I seriously doubt that."

Vivian shot up from the couch. "I hate him, I told you. He has cost me so much. If he were here in front of me, I'd rip his throat out!" She hurled the snifter against the wall. It shattered and for a fleeting second, she thought she saw the face of Ralston Mead appear in the almost translucent metal.

"Rip his throat out? Why, Vivian, you had him in your grasp on the ship and you helped him escape. That is far from ripping out his precious throat." Cobalt leaned against her and she closed her eyes in fear. "You feel guilty about something. Why do you feel guilt, Vivian? Was it the deputy you killed in Lakeside?"

Vivian tried to pull away from him, but he took her arm in his powerful grasp. "The only thing I can think of, my dear Vivian, that would keep you from killing Steel is if you are enamored with him. Or, you plan on using him to defeat me."

"I was told to leave him out of this." Vivian whispered hoarsely. Her demons giggled within and she was on the verge of losing control. She hated Steel more than ever. "I am to defeat you alone. Without the help of Jonathan Steel."

Cobalt whirled her around. His lavender eyes glowed with hypnotic power. "And, the other man you saved? Who was he?"

"Arthur Knight." Vivian said.

"Really?" Cobalt released her in shock. He stepped away, momentarily speechless. He rubbed his jaw and moved to sit on the couch, placing his empty brandy snifter on the edge of the table. It fell when he released it and rolled across the carpeted floor underneath the couch. "My, that is a name from my past. He is dead. I saw to it, Vivian. He is dead. His usefulness to me was over." Cobalt gasped and looked at Vivian. "That explains it. The children had a secret guardian." He smiled and sat back into the folds of the couch. "Oh, what a clever man. It takes a lot to fool me, Vivian. Doesn't he have a son?"

"Josh Knight. Cephas Lawrence's nephew. He and Steel are the boy's guardians." Vivian said trying to wrap her head around this development. Cobalt had been connected to Josh's father? What an intricate web circumstance was weaving around her.

"Now, there is an interesting name from the past. Cephas Lawrence. Arthur Knight. Jonathan Steel. Do they actually think they can stop me?" Cobalt glared into her eyes. "Steel and Knight are now entombed in that cave. I think we have seen the last of them. And, when I have acquired the three pieces of the artifact, no one will be able to stand against me. Especially Cephas Lawrence."

"One of these three pieces is in the crate, isn't it? Is it this Fallen Throne you spoke of?"

"No." Cobalt stood up and walked over to the window onto the clouds. "The Infinity Disc, Vivian. 'Upon the edge of Infinity, the fallen throne resides.'"

"Two components, then? The throne and the disc? What is the third?" Vivian asked.

"I have said enough. I will reveal no more."

The ring pulsed and Vivian covered it with a hand. The Grimvox was more revealing than Cobalt. "I need to powder my nose, honey. Restroom?"

Cobalt stared out the window deep in thought. He motioned to a doorway at the end of the room. "Just inside."

Vivian rose slowly and tried to hide her growing disorientation. The room was fading from the edges inward and she was being pulled into another of Cobalt's stories courtesy of the Grimvox. The restroom door slid aside and closed behind just as she fell through space, plummeting toward a tiny island.

```
Grimvox recall: Subject Icarus
Vol: -256:11234:10:00:03
Crete: Labyrinth of Daedalus
```

Kaballis waited patiently for the young man and his father to finish their argument. He pulled his white robe around his abdomen and secured it with a golden sash. Already, he had subdued the labyrinth guard with a supernatural stupor. Again. The man was so gullible. Every day, he had fallen into a deep sleep and this allowed Kaballis to enter the labyrinth without fear of interruption.

"You should listen to Donnatis. He tells me of the foolishness of these wings. Men are not meant to fly!" The father's voice echoed across the central chamber of the labyrinth. He glared at his son and stalked across the stony ground. Kaballis paused just inside the corridor of the massive labyrinth and a figure in a pale green robe stepped from behind a nearby boulder. Donnatis' hair was as black as night and cut close to his scalp. A beard encircled his face beneath glowing eyes.

"You!" Kaballis hissed at the sight of his old adversary. "Turning father against son. Your Master must be unhappy."

Donnatis unfurled white wings from beneath the robe. "I am saving the young man's life. You would fill his head with Lucifer's lies."

"One day, man will fly, Donnatis." Kaballis smiled. "I might as well be the one to encourage the discovery."

"Leave the boy alone." Donnatis stretched out his wings.

"He is of sufficient age." Kaballis said. "He can make the Choice. You have had your say. Leave us."

Donnatis' wings lowered and disappeared beneath the robe. He faded into the shadows and Kaballis stepped into the open. The young man turned in his direction and smiled.

"Kaballis! I have finished the wings!"

Kaballis glanced up out of the rocky enclosure to the bright blue sky above them. Somewhere up there waited his prize and this young man was the key. "I take it your father still disapproves?"

"He does not wish for me to remain a prisoner in this labyrinth with him. The only escape is upward." The young man glanced after his departing father and shrugged. "He thinks it is folly to consider that a man can fly. But, you were right! I

studied the bird's wings. I finished the wooden frame. And, I've used the flesh of the sea otters to cover the wings."

"What of your father's wings?"

"He refuses to try them." The boy's face lit up with hope. "But, you could try them out for him, Kaballis. If we wait until night falls, the guards will never see us take to the sky. I can fly with my wings and you can test my father's. Once you return to the labyrinth he can follow me and we can escape."

Kaballis smiled and walked across the rocky ground to the young man. "I would be glad to try out your father's wings, Icarus. We shall wait until sunset."

"Where is your father?" Kaballis followed Icarus around the rocky point on the far side of the labyrinth. The young man carried a flaming torch.

"I gave him some hyssop in his evening wine. He is asleep." Icarus led him up a steep pathway to the top of a small hill. At the apex, a flat expanse of rock provided a workplace for the construction of the wings. Two pair of black wings lay on boulders.

The setting sun was below the horizon and the nearby city was alive with torchlight. It was a beautiful sight but not as compelling as the stars that gleamed above. Kaballis had studied the moving points of light in the night sky and knew that soon his prize would pass directly overhead.

"You have done good work, Icarus. What holds the skins together?"

"I have stitched them together with strong cord and sealed the cracks and crevices with wax until the surface was smooth just as you said it should be." Icarus placed the torch in a crack in the boulders and began to strap himself into one set of the wings.

"It is a good thing we are not flying in the daytime or the sun might melt the wax." Kaballis stepped into the frame of his wings. "I assume that we are going to launch ourselves off into the wind?"

"Just as you suggested. Like the gulls soaring on the

breeze, we will glide away from this prison." Icarus lifted the wings off the boulders and walked clear of them.

"We will need to fly higher than this small mountain, Icarus, if I am to return for your father." Kaballis stepped away from the boulders and began to summon his power to bring the winds toward them. "Would you like to soar high above the world and look down upon its beauty?"

Icarus' face filled with the excitement; the promise of the unknown. His face was alive with the Choice! It was so near! To choose to do the unthinkable! To choose to defy the laws of nature! To choose to leave his father and reason and rationality behind. Such would be his undoing.

"Yes! Kaballis, let us fly!"

A great wind blew across the peak of the small mountain and lifted the wings beneath Kaballis. Now was the time! "Run, Icarus! Fling yourself into the hands of destiny! Fly as no man has ever flown!"

Icarus ran toward the edge of the escarpment and disappeared from view. Kaballis lifted himself with his own power and soared after the young man. He extended his own wings into the substance of the primitive wings made by Icarus.

Once over the edge of the escarpment, he saw Icarus soaring beneath him. He dipped his wings and came below the young man. Icarus' face was filled with joy and excitement. "Kaballis, this is incredible."

"Shall we soar higher? If we fly high enough, we will reach the light of the setting sun."

Icarus' smiled faded. "And, what of the heat on the wax?"

Kaballis smiled as the wind buffeted around him. "The heat of the setting sun will not be sufficient to melt the wax. My young man, we must see this world as the gods do. Let us fly higher."

Icarus nodded as the smile returned to his face. "Yes!"

With the Choice now made, Kaballis reached out and created an updraft. The wind swelled beneath them and they soared higher. He directed their path over the city. Beneath them, torchlights illuminated the streets and houses of the city. Icarus moaned in wonder and pleasure.

"It is so beautiful, Kaballis. I could stay up here forever."

As they flew higher, Kaballis pulled oxygen into the sphere of air around them. They breached the darkness and entered into the light of the fading sun. Higher and higher they soared until the earth was a curved line in the distance and the sun was a pinpoint of fading light.

"I have never imagined such a thing." Icarus whispered as they circled in an eddy of wind. "All the world below me and all the sky above me. I am a god, Kaballis."

"Yes, you are, Icarus. You have conquered this world."

Kaballis glanced into the distance and saw the gleaming pinpoint of light coursing toward them. It was still far away but it was beyond his reach. Already, he could sense the Barrier pushing against him. He could go no farther. But, he could propel the young man. He could push him with great power and as a mortal, Icarus would pass easily through the Barrier.

"Icarus, see that approaching star?"

Icarus' eyes reflected the rays of the disappearing sun as he studied the sky. "Yes! What is it?"

"It is the heart of a god. Imagine if you could hold it in your hands. You would become a powerful god among men." Kaballis circled around Icarus. The young man was no longer aware that he was hovering motionless in the thin atmosphere of Earth. His heart and soul were consumed with the possibility of becoming God. Had it not always been so with men?

"How do I get it, Kaballis?"

"I can fly quickly beneath you and create a funnel of wind that will push you higher so you can catch the soaring star. Are you willing?"

Icarus' face darkened as the sun disappeared and all was blackness around them. But, the approaching light illuminated the young man's eager face. "Yes! Yes!"

Kaballis gathered his power and energy and tightened the cloud of air around the young man. With a mighty push, he shoved Icarus upward toward the sky; up past the Barrier; up to the heavens that were denied him.

"You fool!"

Kaballis jerked his gaze away from Icarus. Donnatis

hovered in the sky beside him. His white, feathery wings were extended. "Away from me, Donnatis. The man has made his Choice."

"You think you can pluck the Bloodstone from the night sky?"

"No, but the mortal can. Look!"

Already, Icarus was soaring higher and higher. But, suddenly, he seemed to pause, gasping for breath. The air cloud was gone!

"Your powers are restricted by the Barrier! The mortal will die!"

Kaballis launched himself away from Donnatis and ran into a wall of dark energy. The pain was excruciating and he was thrown back toward the earth in agony. Above him, the mortal collided with the bright light and it fell toward him. Perhaps he could catch it as it fell! Kaballis tried to gather his wings and his power, but he was paralyzed by the collision with the Barrier.

The great light fell far away from him, plummeting to earth in a fiery ball. Kaballis continued to fall and the air around him warmed with friction. He glanced up at Donnatis hovering at the Barrier as Icarus' body plummeted past him.

"I cannot help him, Kaballis. His Choice has condemned him. You are to blame." Donnatis disappeared from sight.

Kaballis became hotter and hotter and he watched Icarus' wings ignite with heat as the young man fell to earth. Kaballis' power returned to his wings. He caught the wind in the upper atmosphere and followed Icarus' body to the earth. The remnants of Icarus' wings took his body into a spiral and his descent slowed. Whirling madly, the body of Icarus and the blackened wings fell back into the labyrinth. Icarus' father waited with a torch held aloft at the apex of the small mountain. He hurried down the path to the foot of the escarpment. He tossed away the torch and knelt over the charred remains of his son. Kaballis alighted in the darkness behind the two.

"My son, I told you the heat of the sun would melt the wax! What have you done? What have you done?" Kaballis watched with disdain as the father clutched his bleeding,

charred son to his chest. The foolish mortal had failed him. But, he had the solace that somewhere, the boy's soul was now a part of his master's dark kingdom. He launched himself into the night as his full power returned. This night had not been a total loss. Somewhere on this planet, the Bloodstone had fallen. And, he would find it!

Vivian gasped as the restroom returned. She looked at her face in the mirror. How did this 'Bloodstone' figure into Cobalt's plans? Was it the third and final artifact? She straightened her hair and touched up her lipstick. She smiled and placed a hand on her hip. She was in control again.

Vivian left the restroom and sauntered across the lounge. She folded a leg beneath her, tightening the hem of her skirt against her thighs as she settled into her chair. Cobalt had retrieved his glass and refilled it with brandy.

"When you were on Dr. Sebastian's show you mentioned an artifact that fell from heaven, the Jupiter Stone. Is that the third part of your precious artifact?" She said quietly.

"Well, aren't you attentive." Cobalt cocked his head. "It is a stone with great power, Vivian. Without it, I cannot unite the three parts. But, it is lost in antiquity."

Vivian tried to act coy. "You know my husband, Robert Ketrick, collected artifacts."

"I have been through his things, Vivian. The Jupiter Stone is not at the FBI warehouse."

"I recall reading something about it in his notes. It fell from the sky like a meteor, right, honey child?"

Cobalt regarded her with renewed interest. "Yes, it did. The last mention of the object in history was at the temple of Artemis where it received its name. Alexander the Great had tried to take the thing and it disappeared."

Vivian glanced at the ring. The stone was dark. She was sure Cobalt had been there with Alexander the Great. Kaballis? Could that be his secret name?

"Who took the stone from little old Alexander the Great?" She asked.

"My adversary. One of the messengers of the light, you might say. He has thwarted me throughout the ages."

Vivian swallowed. Could the stone be in Ketrick's hidden things? If she could get her hands on the stone, she could stop Cobalt. Or, perhaps help him to succeed?

"Help him, Vivian." Vitreo whispered behind her eyes. "Give him what he wants and learn how to control what he seeks to control. When the time comes, we will strike."

Vivian stood up and slid onto the couch beside Cobalt. Her hip snugged up against his and she took the snifter from his hand. She drew a deep breath and swallowed a sip of brandy. She traced her finger down her throat to the base of her neck. "Let me tell you a story. Once upon a time, a man who was obsessed with collecting arcane objects of art sold his soul for a painting. It was no ordinary painting. No, this painting was a self portrait commissioned by the man when he was quite young and stolen by a woman of power and wealth. Unknown to this man, the painting had been commissioned by her and bewitched at the request of the woman so that each and every night, the image in the painting came to life and left its canvas to join her in endless pleasures only to have to return to the lifeless canvas the next morning." She sipped another swallow of brandy. "Now, the ancient crone had died and the elderly man had reclaimed the painting from her estate sale. He hung the painting above the fireplace in his bedroom and each night as he drifted off to sleep he longed to be the young man in the painting. He pleaded with dark forces and powers and principalities for immortality. One morning he awoke and stared down at his ancient, dead body. He found himself trapped in the painting, frozen in cold, sterile immobility. As his children and grandchildren took away his most prize possessions, he had to watch in silent agony. Finally, he stared helplessly into the eyes of his oldest son and saw in that gaze hatred and revulsion. Hoping the painting would at least be hung in his son's home he was disappointed when the painting was taken up into the attic and sealed inside a hidden compartment along with many of his other collected objects of evil. Hidden away from the outside world, he was tormented by the evil spirits of his other collections for all of eternity."

Cobalt took the glass from her and nodded. "You surprise me, Vivian. I had no idea you were such a good storyteller."

"It's more than just a story. Anthony, honey, not all of Ketrick's artifacts are in the warehouse."

"What?" Cobalt glanced at her. "Tell me more."

"He has a house. By the lake. In Louisiana." She milked the moment and took back her brandy. "There is a cellar in which he placed many of his most evil possessions in compartments in the walls." She leaned toward him and giggled. "Why, Anthony dear, your precious little old Jupiter Stone might be tucked away in his basement."

"Well, aren't you full of surprises, Vivian. Then, by all means, let us pay a visit to Louisiana."

"Who knows." Vivian leaned against him. "My poor, late husband might have that Jupiter Stone in his collection."

"Are you flirting with me, Vivian?"

"Why not? I've got nothing better to do."

Cobalt lifted her hand and kissed it with cold lips and studied her with naked lust. "You would be a spirited and memorable diversion. But, that is exactly what you are planning, isn't it? You are trying to divert me from my most important task." He pushed her hand aside. "It is a pity. I could teach you things you have not even imagined. I could take you to the edge of death and ecstasy in one breath and bring you back from the brink of oblivion. I have lived a thousand life times and in each, I have known pleasures long forgotten by men. To share intimacy with me, my dear Vivian, would equal death as you cannot conceive. Remember that."

"You can't blame a girl for trying."

"Now." Cobalt's eyes glittered with interest. "Tell me more about these hidden treasures."

Chapter 17

The ride back to the house had been long and tense. Cephas hadn't said a word and Theo mumbled most of the way to himself. Josh decided to go for an evening swim in the backyard pool to clear his head. He floated beneath endless stars, the sky a fantastic sweep of heavenly beauty.

"It's the next demon, isn't it?" He said out loud to no one in particular. "You guys are trying to hide it from me, but I know. I'm not stupid. Dude, I'm vice president."

He rolled over and began to swim, drawing himself through the water, feeling his muscles burn with the exertion, working away the tension. He was on his twentieth lap when he heard the gate creak open. Uncle Cephas was probably coming out by the pool to smoke his pipe. Josh pulled himself up at the edge of the pool and glanced over at the tables. No Cephas. He looked around the pool deck. It was empty. But, the gate that led out to the side yard and down to the lake was wide open. Must have been the wind.

Josh turned his back to the pool deck and let his legs drift out in front of them. They were burning with the exertion of his swimming. Maybe he could make the swim team. No, they didn't have swimming at the high school he would attend. They had Lacrosse. Should he try out for the team? Surely the girls went for Lacrosse dudes.

Something grabbed his hair and painfully pushed his head down under the water. Josh screamed in agony and he sucked water into his mouth and airway. He started coughing, trying his best to keep from inhaling more water. He reached up to grab the hands that held him under. They were hard and lean. He blinked up through the water and looked into the face of a man from his past. Bile grinned at him and held him under until darkness descended. But the thing that alarmed him most was the spiral that pulsed around the man's right eye.

Cephas watched Bile drag the body of Josh across the basement floor and toss him into a chair. He squatted behind the boy and began to tie his hands behind him. "Don't worry, old man. He's alive."

"If you hurt him." Cephas said through his swollen lip. Bile had caught them unawares after Josh had gone out for a swim. Cephas was tied in his desk chair in the basement. Theo was slumped at the foot of the stairs.

Vivian sauntered down the stairs and kicked Theo in the ribs. "Never get tired of tazing this one. So, Cephas, honey child, you should have stayed in Florida."

"What do you want?"

"Something special. Something hidden in these walls."

Cephas gasped and tried to keep his gaze from wandering to the wall where *it* was hidden. "What are you talking about?" Before he left for Florida, he had found the thing in the wall using the Metastone from the eleventh demon, Lynn Alba. Behind the innocent appearing wall lived a formidable force. If he could somehow activate it before they found it, there might be hope for them.

Vivian drew closer and gazed into his eyes. Her breath smelled of alcohol. "The Jupiter Stone? Temple of Artemis? Where is it?"

Cephas looked away. His heart raced. They were looking for something else. "I don't know what you're talking about."

Vivian leaned close to his ear. "Bile and your old nemesis are allies, but you know that. Cobalt doesn't know. He's upstairs on his cell phone and he'll be down here in a minute. Imagine what will happen when two of the most powerful demons in the world go after Josh." She leaned back and put a finger to his lips. "If you tell Cobalt about number thirteen, you'll die. But, if you help me with the Jupiter Stone, I'll see to it that number thirteen protects you from Cobalt."

Anthony Cobalt? He was here? Cephas tried to swallow but his mouth was dry. If Cobalt was involved, things just went from bad to worse. "Thirteen? What do you mean?"

"Cephas, me lad!" Bile said with a heavy Irish accent. "How soon you would be forgetting me!"

Vivian glanced at Bile. "What's with the accent?"

"Why, Cephas and I are old friends. I helped him get rid of number eleven. Didn't I, old chap?" Bile tapped his right cheek bone where the spiral tattoo began its journey around his eye. "Of course, I wasn't using this dunderhead as my host, but any old human soul will do." His eye widened and he glanced over Cephas' head toward the stairs. "Oh, ho, ho! He's coming. Time to take a break. Wouldn't want Tony sensing my presence." The tattoo faded from Bile's eye. Bile shook his head and looked around the room in confusion.

"Cobalt is the tenth demon, isn't he?" Cephas mumbled.

"Yes. And, I have no idea what he is up to." Vivian said.

"But, you are going to stop him?"

Vivian glanced over her shoulder and a shadow appeared at the top of the stairs. "But I can't stop him unless I know his plans. Be smart, old man. Give him what he wants."

Anthony Cobalt moved down the stairs with incredible grace. "It would seem my sensors have picked up activity on the sun. Vivian, time is growing short." He paused and grimaced at Theo's body. He stepped around him and stood before Cephas.

"Dr. Cephas Lawrence. I once dated a young woman who worked at the library in New York City. My previous host was researching genetics at NYU. He was a bit of a recluse. Much like you." He smiled and raised a hairless eyebrow. "Oh, the things she told me. The books she had tucked away in special hiding places. Books that you wanted no one to see, it seemed. You do have your secrets don't you?" He shrugged. "Well, she died a happy girl. What can I say."

"I thought it was a mugging." Cephas said through clenched teeth. "It was you?"

"Before she died, she told me many things about you. You didn't realize what an eavesdropper she was. All the little things she wrote down in her diary." He nodded. "Yes, Dr. Lawrence, too bad your friends don't know all of your secrets." He glanced at the walls. "Ah, and the things Robert Ketrick hid from the world. He once called me a charlatan and yet here I am rifling through his secrets while his soul burns in hell." Cobalt reached out a hand toward the wall nearest to him and closed his eyes. "And, here we have more hidden secrets. Vivian was correct, I see. I sense an overwhelming galaxy of evil behind this facade. Oh my, the things that are hidden here! Now, Dr. Lawrence, we can decimate this cellar and bring down this house on you and your friends, but Vivian seems to think we need to keep you alive. So, I'll give you one chance to reveal to me the location of the Jupiter Stone."

Cobalt moved across the room to Josh. The boy was slumped in the chair and Bile had receded into the shadows behind him. Cobalt lifted the boy's head and studied his face. Josh opened his eyes and glared at Cobalt.

"Dude, who are you? Captain Milkshake?"

"Super heroes, the bane of this generation. No, my son, I am your worst nightmare." Cobalt smiled.

"Bro, I've seen worse."

"I shall change all of that. Has your uncle told you the good news?"

Cephas shook his head. "No."

"What?" Cobalt's lavender eyes danced with mischief. "Uncle Cephas hasn't told you that your father is not dead?"

"Dude, my father died when I was eleven." Josh looked at Cephas.

"Josh, my father beat me every night with a leather strap embedded with metal studs. Want to know why? Because I was different. No hair! But, the real reason was because I was smart. I made the best grades in school. In South Africa, if you were smart you attracted unwanted attention from the government and my father sold guns to the rebels. He had to make a choice. Keep me under his control or face the firing squad. He was a fiend and a liar, Josh." Cobalt squatted down in front of Josh. "You see we do have something in common. Our fathers. Only, in this case, my father is long dead. And, it wasn't the firing squad. I fed his liver to our hogs. Now, as painful as it may seem, you must realize your uncle is a liar. Just like your father. Just like Jonathan Steel. Isn't that right, Dr. Lawrence?"

"Stop it, Cobalt."

Cobalt kept his gaze riveted on Josh. "Then tell me what I want to know."

"What is he talking about?" Josh asked

Cephas looked away. "I'm sorry, Josh."

"Wait! My father is alive?" Josh's voice broke. "How long have you known?"

"Jonathan told me last night. Your father was at the school looking for these children." Cephas whispered.

"Jonathan knew?" Josh's faced reddened. "Dude, did like everybody know but me?"

Cobalt rubbed Josh's wet hair and slowly stood up. "Poor Josh. Always the last to find out."

Josh's head slumped on his chest and he gazed at the floor as tears dripped onto his chest. Cobalt shrugged.

"Well, there is another way, Dr. Lawrence. Did you know that the biggest lie out there in the field of energy research is that we are running out of petroleum products? We were supposed to run out by the year 2000. No, the truth lies in an industrial complex with powerful political pull that can create any narrative it chooses. Every election cycle we learn of a new form of 'renewable' energy. Now, you and I both know that energy is never renewable. This universe is running down; succumbing to the inevitability of entropy. But, if there were to be some help from beyond nature, let's say from the realm of the supernatural, one might find endless supplies of energy." He reached into his coat pocket and removed a small yellow sphere. He held it up to the light and it glowed with yellow light. "The Sunstone, Dr. Lawrence. My latest creation. Each tiny stone contains crystallized energy. An impossibility according to the laws of physics. But, we both know that our most educated scientists have made the mistake of dismissing what they cannot see with their senses. Too bad. If they had just paid more attention to the supernatural, why this world would be so different." He leaned over Josh and grabbed his head. Josh looked up and tried to pull away but Cobalt shoved his head against his chest and placed the sunstone in the opposite ear.

"Imagine what would happen if that Sunstone were to release all of its energy at one time? It would be like a blast furnace sprang to life inside your nephew's brain."

Cephas strained against his restraints. Josh tried to move his head against Cobalt's fierce grip. "There is one physical parameter I could not change, Dr. Lawrence. The Sunstone is triggered by the ordinary process of heat exchange. I can assure you that Josh's metabolism will be enough to start the inevitable process. And, Josh, the more you struggle, the hotter your head will become. Now, the Jupiter Stone?"

Cephas was breathing hard and sweat poured down his forehead. He glanced over at the wall and the thing that was hidden behind it. "I will tell you. You must understand that I found out about this right before we left for Florida. It is tied to someone I love, someone I have lost."

"Molly?" Cobalt said. Molly was the mother of the young girl who had grown up to become the eleventh demon. He had thought Molly was dead.

Cephas glared at Cobalt. "What do you know of Molly?"

"She is alive and well, I can tell you that." Cobalt said nonchalantly while Josh struggled in his grasp. "And, she is helping an old friend of yours in Europe."

Cold ice ran in Cephas' veins. "Europe? Who?" Cephas strained at the bindings.

"Reveal the Jupiter Stone and I will tell you all."

Cephas looked away from Cobalt. Molly was alive and somewhere in Europe? He had suspected as much from what her daughter had told him when she was no longer in the possession of the eleventh demon. The thing in the wall had been a surprise discovery and he had planned on using it to locate Molly. Just not this soon. "I will need to be freed of these bonds."

Cobalt bumped Josh's head with the palm of his hand. The Sunstone popped out of his ear and Cobalt caught it before it hit the floor. He gestured to Bile. "Very well, but it you try anything foolish, Josh will have one of these in each ear."

Bile stepped out of the shadows and began to untie Cephas' bindings. He leaned forward and whispered into his ear. "He'll never let you see Molly, old man."

Cephas stiffened as the bindings were released. He rubbed his wrists and stood up. "There is a necklace in the top drawer of my desk. It has a green stone on a silver chain." He pointed to the desk. "It will lead me to the stone."

"Vivian, get this necklace." Cobalt still held Josh's head against his chest.

Vivian hurried across the room. She opened the desk drawer and withdrew the necklace. As it hung on its silver chain and twirled in the light, it changed from green to red.

"Is this it?"

"Yes. But, it will only work for me. It will not work for someone in league with evil. You will have to let me use it."

Vivian glanced at Cobalt and he nodded. She handed Cephas the necklace. He carefully placed it around his neck. The Metastone fell over his heart and he sensed it quicken with evil; with power; with desire. He was instantly at war with the thing. He saw Molly walking on the streets of a European city and he wanted Cobalt to die. He saw his old friend, Father James falling beneath the attack of the demons in the abandoned asylum so many years ago. Cephas closed his eyes and fought for control, praying to God for strength to overcome this temptation. He filled his mind with images of Vivian and Cobalt and the thing moved behind the wall.

"Kill them." He whispered.

The wall exploded behind Cobalt, throwing stone and sheetrock into the air. Cobalt was tossed across the room like a scarecrow. The thing came out of the wall, all stone and dirt and rusted metal and dead glass; bits and pieces of trash and debris fused into a roughly human figure seven feet tall. Its stumpish head was short and squat and two glowing red eyes fixed on Cephas. Its flesh was earth and rock. Its heart was the red glowing Bloodstone encased in a net of meteor rock.

"Golem obeys, master." It growled and it headed for Vivian. Cephas stumbled over to Josh and began to untie him.

"Dude, what have you done?" He said.

The Golem swatted Bile across the face and Bile was flung across the room and into the corner behind unopened crates. Golem turned to Vivian, cornered against the wall. Now, she would pay for all the evil she had brought upon them. Now, she would die and Cephas reveled in that outcome. His face twisted in a leer and he screamed, "Kill the witch!"

The ring warmed on Vivian's finger even as the Golem creature hunched over her. Its horrific face was filled with fine bits of broken glass and twisted metal and frozen earth. It glared at her with inhuman eyes. She put out her hand to push the thing away and touched the Jupiter Stone in its chest. Her ring pulsed with life and the Golem lurched as the stone netting around the Jupiter Stone shattered. It fell away to reveal the glowing red Bloodstone. Golem straightened and blinked. "I obey you, master."

A new sense of death and destruction filled her head. She wanted nothing more than to kill, to maim, to destroy all that stood in her way. Vitreo screamed with delight. Her other demons swelled with sudden power. Cobalt climbed out from under the debris and she smiled in wicked delight. "Kill him."

Golem turned to face Cobalt and unleashed a powerful blow. But the blow never fell because the hulking mass of Theophilus Nosmo King arose behind Cobalt and drove him across the room with a body slam. Cobalt bounced against the computer desk and fell over into the space behind it. "It's time for me to kick some demon butt!" Theo screamed and he glared at Vivian. "Starting with you."

Vivian nodded at Golem. "No, starting with you."

Golem crouched and jumped across the open space. Theo

met the thing with all of his force and the sound of their collision shook the foundations of the house. They crashed into the wall where Golem had been hiding and on through the other side of the separating wall into the center of the cellar. Golem tossed aside splintered and broken crates and drove a stone fist into Theo's stomach. Theo was driven up into the air and over the crates.

Theo rolled through the debris and grabbed a handful of broken concrete and hurled it at the Golem. The Golem opened wide its arms and the concrete struck its abdomen. But, instead of reeling from the blow, the creature absorbed the concrete and enlarged.

"That ain't right!" Theo mumbled through swollen lips. The Golem swept aside the crates between it and Theo and grabbed the man by his neck. Theo struggled against the power of the creature.

"Stop!" Cephas screamed. "You have what you want. Leave Theo alone."

Cobalt rose from behind the desk on black wings of crystalline darkness. The tattered remains of his shirt drifted to the ground. "Cease this attack, Golem."

Vivian glanced at her glowing ring and hid it behind her. She couldn't let Cobalt know about the ring. "Golem, stop!" She shouted.

Golem froze and Cobalt glared at her. "How did you do that?"

"I touched the Jupiter Stone and I guess it responded to my demons. I gained control over it."

Cobalt tore his gaze away from her and sailed across the room on crystalline black wings and landed in front of Cephas. He jerked the necklace from around his neck and held the Metastone in his hands. "Deal's off, Dr. Lawrence. You betrayed me." He glanced at Golem. "Vivian, tell Golem to kill them all."

Before Vivian could speak the flesh of Golem's arms began to crumble; flecks of stone and metal tumbled away and Golem disintegrated before their eyes, collapsing into a pile of dirt and debris. The Metastone crumbled into black ash in Cobalt's hands.

"What happened?" Cobalt's wings folded out of sight.

"The Bloodstone, or Jupiter Stone as you called it, is far more powerful than the Metastone." Cephas breathed a prayer of thanks. He had no idea why the stone crumbled. "I would imagine once the Bloodstone was freed from its meteor casing, its power destroyed the Metastone. The Bloodstone is the master of all such talismans. The two cannot co-exist. The Bloodstone neutralized the Metastone and once the stone that controlled Golem was destroyed, Golem was destroyed with it."

"Well, it no longer matters." Cobalt tossed the medallion into the pile of Golem's ashes. He picked up the huge Bloodstone. It was the size of a small melon but shaped like a drop of blood. Its deep, crimson interior glowed with power. "Looks like you have won a reprieve, Dr. Lawrence. I now have the Bloodstone in my possession. Vivian instruct your underling to kill these three and let's get on with our quest."

"Wait!" Vivian said as the spiral began to appear around Bile's eye. "You need the third item, right sweetie? The Fallen Throne? Renee Miller is searching for the throne. If anyone knows where it will be, she does. These three men are friends with Jonathan Steel who is helping Renee find the children. They are still of use to us." She pointed at Cephas. "Get the old man to call Steel. Send Steel to find the throne or kill his friends. Simple. Clean. You said time was running out on you."

"But, Steel is dead."

"He is quite alive." Cephas said. "He escaped from an underground cave and is on his way back to Florida right now."

Cobalt's eyes glowed with demonic power and he screamed in an unintelligible language. He stalked around the room, deep in thought. He glanced through the broken wall at Theo struggling out of the broken remains of a crate. "Old man, I will let you live for now. But, you must make the call to Jonathan Steel. I have an offer he won't refuse." October 13, 2015

Chapter 18

Steel hated waiting. Patience was not one of his virtues. He inhaled the smoky air and coughed. His chest burned from the tainted air in the chamber. He paced and watched Jessie pick up the pieces of his friend's body and stack them in an empty equipment chest.

Sebastian stood in front of the massive mural and whispered to Arthur Knight. He drew another lung full of air. How long would it last before they opened the doors to the outside?

"Cassandra, hopefully your sister will get here before we suffocate." He said.

"Cassie." She looked at him. Soot and sand streaked her face. "Please call me Cassie."

"Well, Cassie," Steel crossed his arms over his bare chest. "I've been in the dark since I met a talking dog in an attic and I need some answers. What is this? And, don't go spouting riddles."

Metal creaked as Knight sat on the equipment case. "Tell him everything, Cassie."

"Everything?" Cassie bit her thumbnail.

"About the artifact, I mean." Knight said.

Cassie nodded and reached out a hand to touch the stone wall. "One of Renee's theories claimed the disc might have been secreted away by a Jesuit order. The Jesuits must have brought the disc here in the 1400's and built the mission above. History tells us they came to teach the Seri Indians agriculture and to convert them to Christianity. I always had a problem with that premise." She pointed to a scene on another wall.

Native fishermen cast nets into the sea. "You see, the Seri culture was based on a fishing society. Agriculture would never work in this climate. The Jesuits would have tried and given up, moved on. So, the Jesuits stayed for another reason. That is where Renee got really suspicious. This island is a perfect hiding place for something the Jesuits wanted hidden from prying eyes." She turned and gestured to the depression in the floor. "They brought the disc and hid it in this cave."

"Where did this disc come from?" Steel asked.

"No one knows for sure." Renee pulled her keys out of her pocket. She held up the key ring and light glittered on a small piece of crystal. "Look at this. A tiny piece of crystal with carvings at the tip. I got this from a very reliable source. A piece of the Fallen Throne." She pointed to the mural. "See how the throne is depicted in a pale blue? This mural substantiates the suspicion that it is of total crystalline construction, carved from one huge block of this substance. Do you know what this crystal is made of, Jonathan?"

"No."

"Lonsdaleite. It forms when meteorites containing graphite strike the Earth with great heat and stress. It is a form of diamond and is over 50% stronger making it one of the hardest substances on Earth. It is exceedingly rare and yet, if legend is correct, an entire throne is carved from one huge deposit of Lonsdaleite." Cassie shoved her keys back into her pocket. "And, I had the crystal dated. It is over 100,000 years old! Think about that! An artifact carved from one of the rarest and the hardest substance known to man that is over 100,000 years old! Who carved it? How did they carve it? Was man even around when this throne was made? It's the Holy Grail of archeology, Jonathan." She grinned. "Pardon the pun. There's supposed to be a sound clip after I say that. In my show. A whip cracking. Like the movie?"

"Is the Infinity Disc made of this crystal, too?" Steel sighed.

Cassie shook her head and her eyes lit up with excitement. "No, Jonathan. Like the throne, the Infinity Disc is another impossible artifact. Supposedly it is made of a metal untouched by erosion or decay. And, it is rumored to be just as old as the throne! It was made long before man could work metal."

Steel closed his eyes. He was beginning to have a good idea how such an impossible thing had been made. If man did not have the science, then they had to have come from a supernatural source. He studied the image of the Fallen Throne and the hair stood up on the back of his neck. This thing was evil. He could sense it. The being that stared back at him was strikingly familiar. "What kind of being is on this throne?"

Cassie smiled and returned to the mural. She reached out a finger and touched the rough stone image. "It's a cherubim."

"What? I thought cherubs were fat little angels with wings?" Steel said.

"No, no! Jonathan, in the book of Ezekiel these types of angels fly around the throne of God. They have four wings and four arms and a head that is composed of four faces. Their job is to magnify and glorify the power of God. The power of God, Jonathan. Imagine sitting on this throne and having at your disposal the power of God?"

Sebastian blinked and her thoughts were obviously far away. "Jonathan, this is not the angel sitting on the throne. This cherubim IS the throne. See, two upswept wings. Two down swept wings that curl about its feet. One set of arms holds the sword above its head. One set of arms rest on its legs making the arm rests of the throne. The chest is covered with a breastplate and that image of a red stone on the breast plate is the Bloodstone."

"And the head. It seems to be some kind of mask?"

"There is a large stone that rests behind the mask and makes the 'head' of the cherubim."

"What stone?"

"The Bloodstone." Knight said.

Steel glanced at him. His gaze was averted from the mural. "The two of you mentioned that earlier. Tell me more. Now."

Knight looked at him and his face was ashen and pale. "The Bloodstone is the third piece of the artifact."

"That's not an answer. What is it?" Steel's face warmed with anger. "I told you I'm tired of riddles and games."

"Jonathan," Cassie pulled him over to the edge of the depression. "Look, the disc is similar to an object known as the Lolladoff disc."

"Just a hoax." Knight said.

"What is a lollipop disc?" Steel asked.

"Lolladoff disc. There is a photograph of a disc supposedly discovered in Nepal. The disc contains a spiral with the sun at the center of the disc. Along the spiral is a disc shaped object thought to be an extraterrestrial spacecraft. And, there is a figure suspiciously similar to the typical 'gray' alien seen in UFO encounters."

Steel sighed. "Like the ship that abducted us? Is this where you are going now?"

"We were abducted by a flying saucer." Knight said. "Jonathan, we were probed by alien like creatures."

"Nightflies." Steel hissed.

"Good name, but you can't deny that something is going on here suspiciously identical to a typical reported UFO encounter." Knight got up from the chest and joined them at the edge of the disc depression. "And, if you think about the appearance of these nightflies, they remind you of that." He pointed to the throne. "Cherubim?"

"So, this Lolladoff disc was created by someone abducted by alien angels?" Steel said. "Come on!"

"You've seen stranger, Jonathan. Giant scorpion. Vampires."

"Talking dogs." Steel sighed.

"Okay, how many extra solar planets have we discovered in the past few years?" Cassie asked. "Hundreds and hundreds! And yet, if all of these extra solar planets contain life, then why is there such a profound silence? Why haven't we heard from intelligent life on other worlds?"

"Cassie, there are many reasons." Knight stood up. "There is the limit of traveling faster than the speed of light. The problem of interstellar radiation. The scarcity of stars like our sun."

"Yes, but there is also the problem of our moon." Cassie pointed to the ceiling of the chamber. Stars and planets were painted against a dark blue background. And near the edge hung an image of the Moon.

"The moon is a problem?" Steel asked.

"Do you know how the moon formed? Well, billions of years ago a Mars size planet entered our solar system at just the right speed and just the right angle and collided with a primitive Earth. In the aftermath of that collision, most of the Earth's atmosphere and water were sheered off into space. And, the collider injected heavy metals and radioactive isotopes into the Earth's core. The remnants of the collider than coalesced over time into the Moon."

Cassie's eyes gleamed with enthusiasm. "If there had been no collider, the Earth would have had too thick of an atmosphere and too much water. We would be like Venus, a hot poisonous atmosphere thanks to a runaway greenhouse effect. Jonathan, there are over 600 individual parameters of physics, the universe, the galaxy, this solar system, and this planet that had to occur in just the right order and just the right quantity for life to be possible on planet Earth. These parameters are unique to Earth. We don't see them anywhere else. All those planets we've discovered? Too close or too far from the wrong type of star. Too large or too small. Yes, some of them have water, but that doesn't mean the conditions are right for life."

Cassie pointed to the disc. "The Lolladoff Disc implies that life is abundant throughout the galaxy. But, the statistics just don't hold up. It is very likely that we are the only planet in the entire universe that can support life. Because the conditions that allowed us to exist are so rare, so improbable that Earth is truly the impossible planet."

"Jonathan, most archeologists think the Lolladoff disc is a hoax. But, the Infinity Disc is real! And it is very similar in appearance to the Lolladoff disc." She stepped down into the depression and walked to the center. "A spiral moves around a central raised circle here." She pointed to the center. "Within that circle are stars carved into the metal. No one really knows what lies along the spiral. The few images that exist are so different from each other and the only common element is the central circle and the stars. Oddly, the star pattern matches nothing we can see from Earth with the naked eye. It is said that when someone who is worthy sits upon the throne, the Bloodstone is activated and the disc's central circle opens up to another world."

Silence fell and Steel watched the childish delight on Sebastian's face. He heard Knight breathing heavily beside him. "This is YOUR Holy Grail, isn't it?"

"Yes!" She nodded and tears made tracks through the soot on her face. "I believe there is another world out there teeming with life. And, if that life exists, then something supernatural must have designed it. Imagine the power of such a designer. Imagine what it could do."

"I don't have to imagine." Steel said. "I've seen the evidence for the supernatural and your designer is God. And if God wanted to make another world out there with life, that's none of my business. We have to deal with this world right here, right now."

"Tell him the real reason you are looking." Knight said.

Cassie jerked as if stung. She wiped at the tears. "What?"

"Tell him!" Knight shouted and his voice echoed through the chamber.

"Please, Senorita Sebastian." Jesse said quietly from the balustrade. Steel had forgotten he was here. "Do not continue to displease God. He has punished you enough."

Cassie glanced at Jesse and nodded. Her lips trembled. "Yes. I haven't forgotten about Maria. I will help her, Jesse. I swear."

"And, yourself, Senorita." Jesse whispered.

Steel waited, this time patiently as Sebastian sniffled and fought the tears. "I'm more than a self centered television star, Jonathan. I do this for a reason."

"Tell me."

Sebastian whirled and her gaze was hot and penetrating. She held up her right hand palm out. "This."

Steel raised an eyebrow and started to say something

when he noticed the fine tremor in her hand. He blinked as sudden memories of another woman with a tremor flooded his mind. He was back on the front porch of the missionary house in Lakeside seated by Claire Knight. Her hand was shaking as she told him about her illness. His breathing came hard as the memory took him fully and he looked away.

"What are you telling me?" He managed to say.

"Other legends say the circle, when activated, will heal you of your ailments."

"Senorita Sebastian will heal my wife." Jesse said.

"Si, Jesse." Cassie nodded. "Jonathan, it is said the spiral is in fact, a model of human DNA inscribed on this disc tens of thousands of years ago." She let her hand drop beside her. "Yes, I believe in God. Yes, I believe He is the Designer! I just have some problems with why God would allow innocent people to suffer."

She paced and wiped her face with the back of her hand. "Jonathan, I believe the image of DNA on this disc is the original human DNA created by God. It is perfect and in pristine condition, untouched by thousands of years of mutation and disease." Tears dripped from her chin. "In the Bible, the original humans lived for hundreds of years. I did an analysis once. I compared the decreasing ages of the patriarchs to a curve reflecting what subsequent centuries of increasing genetic mutation would do to age. The curves matched!" She pointed to the empty space at the center of the depression. "Jonathan, when the disc is activated, it will optimize your DNA returning it to its original condition."

"So you want to live hundreds of years?" He asked. "Is that what this is all about?"

"No. I'm sick, Jonathan."

"And, so is Renee." Knight said. "Huntington's Chorea."

"An inherited illness that is autosomal dominant. No matter which of your parents has it, all of the children will get it. It is a progressive degenerative nerve disease that appears in your forties. Both or our parents died from it." Cassie walked out of the depression. "But, Jonathan, Vega doesn't have it! She doesn't have the gene! Renee tested her. She has something else wrong but not that gene!"

"That is why she is searching? I thought she wanted to fix Vega. You're telling me, it's more like she is trying to find out how to fix her own genes?"

"Yes. What happened in Vega to neutralize the Chorea gene? Why did she manage to escape it? And, did whatever happened to her also activate another gene that is causing her illness? Xenoarcheology may hold the answer." Cassie's eyes were wild with fervor.

Steel shook his head and clenched his fists. "So, you finding this artifact has nothing to do with helping Vega. It's all about you."

Cassie seemed to deflate. She stumbled back into the disc depression. "I am who I am, Jonathan."

"Yeah, a self centered con woman." Steel growled.

From a far distance away, a dull thud came against the doorway. Knight spun and ran up the ramp to the door. "Renee is here!" He pressed his ear against the melted metal. "Wait! I hear gunshots. There's a firefight going on in the pit."

Jesse stood up. "It could be the cartel, Senorita."

"Who is fighting who, though? Renee or the Federales?" Cassie hurried to the door.

Steel hopped over the stone balustrade and shoved against the door. "We have to get out of here and help Renee. She's not ready for a fire fight with the cartel."

"Steel, it's melted solid." Sebastian said.

"How can we open it?"

"I could open the door if I had the, uh, artifact the nightflies were looking for." Knight said.

"What?" Steel reached into his pocket, took out the necklace and held it up. The large, blood drop shaped jewel glittered in the light. "Are you talking about this? If so, get us out of here." He handed it to Knight.

Knight gasped as the necklace collapsed in his open hand. "Where did you find this?"

"At the school."

"What?" Sebastian pushed Steel aside and reached for

the necklace. Knight closed his fist and pulled away from her. She swore and ran a hand through her hair. "I can't believe it! Do you know how long I've searched for this? And, you had it all the time?"

"It's just a splinter off the real Bloodstone. And, using it comes with a cost." Knight opened his fist and studied the jewel in his hand. A haunted mask fell across his face and he swallowed. Slowly, he took the chain and placed it around his head. The stone fell across his chest and came to rest over his heart. Instantly, it began to glow from within and Knight gasped, stumbling back and reaching for Steel for support.

"Arthur?"

Sebastian tried to push around Steel and he restrained her. "Leave it alone, Cassie. You've done enough today."

"I'm okay. I'm okay." Knight drew a deep breath and straightened. His breathing slowed and a red light filled his eyes. He faced the deformed doors. "You might want to take cover."

Steel grabbed Cassie by the arm and pulled her down the ramp and behind the stone balustrade. Jessie crouched beside them. Knight breathed slowly in and out, coughing now and then with the thinning air. He slowly raised his hands, closed his eyes and began to speak quietly. His hands began to glow with an orange light and his eyes snapped open. He shoved his hands forward and shouted at the top of his lungs.

"Open!"

The light that filled the chamber was blinding and the walls shook. A huge rumble filled the chamber and bright light flooded in. Stone dust filled the air and Steel hopped over the balustrade. Arthur lay in a heap in the beam of bright light gushing in from outside. Smoke poured in and a lone figure walked out of the haze. Renee Miller wore combat fatigues and carried a huge automatic rifle.

"Where is my sister? It's time for me to kick her butt!"

PART 4
UFO CULT

Most UFO cults share three interrelated beliefs:

1 - Flying saucers are real.
2 - People are in touch with alien intelligences associated with flying saucers.
3 - The messages given by aliens are of immense importance to human beings.

Hugh Ross

Chapter 19

Vega sat up on her bed when she felt the unmistakable thud as the ship settled onto hard ground. She saw movement in the corridor and for a moment she recalled the hideous creature of night and shadow. But, the shadows formed themselves into the more substantial forms of her friends. Altair leaned her head across the threshold. Her hair was as dark as the night sky and her piercing green eyes danced with mischief.

"Hey sleepy head, get up and let's go meet my mother's friends."

Vega shook her head. "I don't know if I want to meet your mother's friends. Some of them are weird."

Altair stepped into the room with her head bowed to keep from striking the doorframe. She wore a pair of jeans and a light, pink sweater with red tennis shoes. "Don't be silly, Vega. My mother's friends are fine."

Vega stood up and her head brushed the ceiling. "Oh, yeah? Well, what about those weird little creatures in the control room? Or, that green eyed monster thing in the hallways?"

Altair planted a hand on her hip. "Vega, you've been dreaming. There aren't any aliens on this ship! It's built by the Enochians and they are all humans."

Vega stepped past her and out into the corridor. The other children were milling around outside their cubicles. She hurried to the door that led into the control room but it was shut and sealed tight. Altair leaned on the wall beside her.

"Where are they, Vega?"

"In there." Vega knocked on the door. "And, Mrs. Donnelly was in this glass tube filled with healing waters."

Altair laughed and the other children joined her. There were six boys and six girls besides Vega. They were all as tall as she was.

"Little green men?" Deneb popped his head in between Vega and Altair. His short, blonde hair was cut close to his scalp and his dark eyes glittered with laughter.

"The claw! The claw!" Tonisha's dark haired head reared up behind Deneb and she dangled her hands like claws.

"Listen, I was in there and--" The door slid open with a sigh and Vega turned. The consoles had been replaced with tables filled with vegetables and fruits. The central glass cylinder was misted and filled with the writhing branches of grape vines covered with red grapes. Magan Celeste turned around from one of the tables carrying a crate of cantaloupes.

"Oh, good. You're all awake. You can help me take the fruits and vegetables out to our new home. Come on in, children and grab a crate."

The children streamed past Vega and she gaped at the room. "This wasn't like this." She stepped over the threshold and towered over Magan Celeste. "Where is Mrs. Donnelly?"

"I'm right here, Vega."

Vega whirled around and saw Mrs. Donnelly standing in the corridor. She rushed through the open door and embraced her, lifting her off her feet. "You're OK. I was so worried. Magan Celeste put you in that tube and --"

"Vega." Mrs. Donnelly slowly pushed herself out of Vega's grasp. "Can we talk about this later?" She winked. Maybe she had not seen the strange men in the control room after all. Or, was Mrs. Donnelly telling her it was their secret? She recalled what Magan Celeste had told the thing in the hallway. She said she had threatened to harm someone in order to get Mrs. Donnelly to cooperate. Could that be true? If she pressed on with her questions, someone might get hurt. And, that meant that Magan Celeste wasn't the loving, kind woman Altair thought her mother to be. Vega stepped back from Mrs. Donnelly and nodded.

"We can talk later." Vega said.

Magan Celeste walked by with the crate of fruit. "I agree. For now, we have to go meet our new friends and see our new home."

Vega followed Mrs. Donnelly and the other children along the corridor to a huge, open doorway. A ramp led down to a platform that must have been some kind of stage. It was like a giant shoebox with the back cut out. She looked behind her as she walked down the ramp. The ship hung in the air and beneath its edge, she saw beyond the back of the open platform a huge expanse of brown desert far beneath them and the sloping curve of a mountainside stretching away beneath the ship. They were perched on the top of a mountain somewhere and the platform was little more than a huge stage set just for them.

A curtain covered the far side of the platform. She followed the rest of the children from the ship, bearing their crates of fruits and vegetables. Off to her side, another ship smaller than her ship appeared in the open back of the platform. It hovered in the heat and a ramp slid from the front edge and landed on the platform. Who else was joining them? And then, she sensed something in the mixture of visible light and infrared and ultraviolet waves of energy whirling all about this surreal scene. Somewhere, someone watched them with tiny eyes and tiny ears, someone hidden from Magan Celeste and the children. She whirled and looked far up into the back corner of the platform, into black metal girders with lights and vents suspended from them. She squinted and saw something gleaming with a subtle pulsing of light. She didn't know what was going on. She had no idea what this platform was all about. But, she did know one thing. They were being watched.

Chapter 20

"So, what is an agent of your seniority doing in this job?"

Special FBI Agent Franklin Ross exhaled his cigarette smoke into the young woman's face and grimaced. "I told you I didn't want to chit chat."

Agent Margie Sculder coughed and waved away the smoke. "You're not supposed to smoke in here."

"You're not supposed to whine." Ross growled. His head hurt. His back hurt. His chest hurt from starting back on the cigarettes. But, he was far away from demons and angels and vampires and arks and, most importantly, Jonathan Steel.

Sculder was a short, perky woman of Asian descent with jet black hair. She wore a smart dress and jacket combination. Ross wore his rumbled white shirt and his stained red tie.

"And, why do you still wear your sunglasses in the dark?" Sculder coughed again.

Ross pushed the sunglasses up on his forehead. "So you can't tell when I'm napping. Although it's a safe bet if your lips are moving, I'm sleeping." He pulled the sunglasses back down and studied the monitors in front of him. "Seen anything yet?"

Sculder sighed. "No demons, if that's what you're worried about."

Ross sat up from his slumped position and tossed his sunglasses on the table in front of him. He looked across the dark room at the woman's face painted in dull light from the surveillance monitors. "Where did you hear about demons?"

Her mouth turned up in a smile. "There's the usual scuttlebutt about vampires and twelve demons. Or, was it thirteen?"

Acid boiled in Ross' stomach and he stubbed out the cigarette butt on the tabletop. "Listen, Mulder, . . ."

"Sculder." She squinted at him.

"Sculder, whatever. You are never to mention demons or angels or vampires or--"

"Jonathan Steel?" She said.

Ross was speechless. He closed his mouth and reached into his pocket for another cigarette. "That's it. I'm done. Get out of here."

"What? I can't just leave!"

He lit up another cigarette and inhaled the warm vapors. "I'm the senior guy on this thing, so get out. Now!"

"You can keep an eye on all six monitors by yourself?"

Ross shrugged. "It's a UFO church. Someone is laundering money through the church. And, it's out in the desert, hundreds of miles from here. All we need is one person to show up with ties to the financial field. We do that, and we've got them nailed." He looked at her and reached for his sunglasses. "And, not a demon in sight. That is why I asked for this position. I needed a vacation from the weird."

"And, a UFO church is not weird?"

"There's a sucker born every minute, Sculder. And, we're looking at a couple hundred of them following Magan Celeste and her Church of the Enochians. Now, go and play some slots. We're in Las Vegas. Come back in two hours and you can give me a break." He glared at her one last time as he pulled on his sunglasses. "And, that's an order."

Sculder slowly stood up. "Hey, there's some activity going on. Right there."

Ross put his cigarette down and leaned into the monitors. The third monitor was taken from the back of a platform used by the church for some of their services. Some kind of metal ramp slid into the frame and a small framed woman made her way down the ramp to the stage.

"This can't be!"

"What?"

Ross pointed to the image. "Vivian Darbonne. Uh, Ketrick. Uh, Wulf. I think one of them is her new last name."

"Who is she?"

Ross shook his head and his stomach churned. "A woman from the past."

Another person appeared on the ramp. Sculder lifted an eyebrow. "And, Anthony Cobalt? What is one of the world's richest men doing at the compound of the Church of the Enochians?"

"We may have found our financial connection." Both Vivian and Cobalt were met by a woman in a flowing white gown. Her bald head gleamed in the sunlight.

"Magan Celeste." Sculder said. "With Cobalt. You've got your connection all right."

More people came down the ramp. Ross' heart raced and the hair stood on end as he saw a man with wispy hair manhandling what had to be two smaller prisoners. He jerked his sunglasses off and tossed them across the room. "No. It can't be! No, no, no, no, no!" One of the men was Bile. From Rudolph Wulf. From the whole thing with the vampires. And, with the Ark of Chaos. Or, was it the Ark of the Demon Rose? This was not good. He slumped in his chair and massaged his face. "Not again!"

Sculder glanced at him and Ross leaned back in his chair. "What? Who are these people, Ross?" She watched the man push a teenage boy and an older man roughly down the ramp. They had plastic ties around their wrists. "Those two men look like prisoners."

"Yeah." Ross sucked on his cigarette. "And, if they are, my past has just caught up with me. The kid is with Jonathan Steel and the old man is the boy's uncle."

Sculder sat back in her chair. "You mean the demon guy?"

Ross opened his eyes. "The very one."

"Well, we better call in a task team."

"And blow two weeks of surveillance? Before we do

anything, I'm calling Jonathan Steel." He picked up his cell phone. "I thought I'd never say that again."

"Ross, you can't do that! It's not agency policy."

"Forget policy. My world just went to hell in a hand basket." He said as he dialed the number. His demons were back!

Chapter 21

Renee Miller crossed the threshold of the blasted metal doors and went straight for Cassie and let the fury and anger boil up inside her.

"You have gone too far this time, Cassie!" She lifted the rifle muzzle and aimed it at her sister's foot. "I ought to shoot you right here. Right now."

"I'm sorry, Renee. I had to have the artifact before you got your hands on it and buried it in an Ingenetics Laboratory." Cassie shoved the barrel away from her.

"And, now the absolute worse person in the universe to have it has taken it right from under your nose."

"It wouldn't have made any difference if you were here. He was unstoppable. I'm only alive because of Jonathan Steel." Cassie stepped closer to Renee. "It was the Infinity Disc, Renee. It was here just as you thought. Brought here by the Jesuits. And, back there is a mural with the Fallen Throne and the Bloodstone. The legend is true, Renee. We finally have proof."

"For your television show?" Renee tried to temper her rising excitement with her sustained anger. "This is not for us. It is for Vega."

"It can be for both of us. But, we have to get the disc back. And, we have to keep anyone from getting the other two pieces of the artifact."

"You're right."

"As usual."

"Don't push it. I just fought off a dozen cartel mercenaries and there are more on the way. We need to get the survivors onto the helicopter and get out of here. Now." She turned to Steel. He was helping a man to his feet. The man's back was to her and he leaned against Steel.

"That took a lot out of me." The man whispered.

Renee recognized the voice. She recognized the hair on the man's head and the shape of his neck. She dropped the rifle and it clattered on the floor of the cave. She grabbed the man and whirled him around.

"Hello, Renee."

She slapped him hard across the face and he stumbled back into Steel. "You! You're supposed to be dead!"

Steel stepped between them. "He just saved our lives, Renee. Take it easy."

"Take it easy?" Renee shouted. Tears stung her eyes. "Arthur Knight is alive and well? You sorry b--"

Knight held up his hands in defense. "Renee, stop. I'm the Guardian. I've been protecting the children. I'm your mercenary."

Renee froze and put a hand to her mouth. "What? You? Well, you didn't do a good job, did you? They're gone. Probably in the hands of Robert by now along with the Infinity Disc."

Knight stepped closer with his hands held up between them. "It's worse, Renee. Anthony Cobalt was here. He took the Infinity Disc."

"Cobalt? No!" Renee gasped. "Can this get any worse?"

"Yes." Knight's hand fell onto her shoulders. "Cobalt has Vega."

Renee's world collapsed around her and she stumbled back and fell onto the dirt floor. She pushed her hand against her mouth and the sobs came. "No! No!"

Cassie knelt beside her. "We'll get her back, Renee. We'll get them all back."

Renee shoved Cassie away and hugged her knees to her chest. Knight knelt before her and held up something red and silver. "I have the Bloodstone, Renee. We can use it."

Renee glared at him. "Get that thing out of my sight! You know what it did to us the last time you used it."

She sobbed for a while until someone sat beside her. An arm draped over her shoulder and Jonathan Steel spoke. "Renee, I don't know what is going on here. All I know is we have to get off this island soon or I will not be able to rescue Vega and the other children." She looked up into his turquoise eyes. "I promised you I would get her back and I will."

A hand appeared before her. "Let's go get our daughter." Arthur Knight said.

Steel tensed beside her. "Did you say our?"

Knight nodded. "Yes. I'm Vega's father."

"Jonathan, we have to go." Renee screamed over the blast of the transport airplane. They had spent three hours in one of Renee's newest military helicopters over the Baja Peninsula until landing in Cabo San Lucas. Now, the transport plane waited to take them back to Pensacola. Steel looked at the sat phone.

"I can't get an answer on Cephas' phone or Josh's phone or Theo's phone."

"Call and check on them when we get to the condo. No signal in the airplane."

Steel shook his head. "Not good enough." He dialed one more number. A warm, female voice answered.

"Hello."

"Liz, this is Jonathan."

"Hello, son. How are things going?" Dr. Liz Washington said. "I've almost completed the dig at the old church site. Wait until you see what I've found."

Steel imagined Liz's dark features and her snowy main of hair and her infectious smile. "Liz, there's trouble."

"And I should be shocked?"

"It's number ten."

Liz drew a deep breath. "So soon?"

"Yeah, so soon is right. I sent Cephas, Josh, and Theo back to the lake house and I can't get an answer on their cell phones. I'm about to get onto an airplane returning to Pensacola." He turned to Renee. "How long before we get there?"

"Four hours." Renee shouted and pointed to the ramp at the read of the transport. "We need to go."

"Four hours, Liz. I need you to send someone out to the house to check and make sure they made it home."

"I can go and check it myself."

"No! There may be danger."

"Jonathan, I'll take care of it. Call me when you land."

"I'm worried, Liz."

"Say a prayer. No need to fear, Mama Liz is here." She laughed and Steel found himself smiling.

"Okay, but be careful." He ended the call and followed Renee up the ramp.

"A helper in time of need." Sebastian glanced at Steel's business card in her hand. "Well, isn't that sweet." She glanced at Renee strapped into one of the side seats in the cargo bay of the transport. "I wish father had been more altruistic."

"Dad never liked our choices." Renee said. "I went into the medicine field."

"To find a cure for your disease?" Steel said.

Renee's mouth fell open and she glared at Cassie. "You told him?"

"I thought we were going to die."

"Renee, it's fine. I understand." Steel said.

"What you don't understand is how Dad felt about me." Cassie tossed the card back into Steel's lap. "When mother got, uh, sick, Dad started looking for a cure anywhere. He had money and influence but it did no good. And, when Mom died I asked for my share of the inheritance. Renee took hers into medicine. I took mine in a different direction."

"Artifacts." Renee said.

"I started out into the world of antiquities at age sixteen. In just eight years, I became fabulously popular and wealthy. Dad never approved. Especially when he saw my museum." Sebastian said. "Renee was there on opening day. Trying to bring peace to the family. And, there was our Dad pulling up in his limo, climbing out, his tall, wiry frame swaying in the warm summer breeze. The museum had been built in an abandoned parking lot."

"'I liked it better as a parking lot.' That was all he said. He climbed back into the car and had his driver pull away leaving me stranded at the podium in front of news crews and dignitaries. At that moment, I hated him more than any other time in his life." Sebastian said quietly.

"Problem was, Cassie, he was right. You're no Indiana Jones." Renee said. "You thought artifacts did not belong in the museum for the world to see. They belonged to you! In your museum! And, it wasn't about money. It was about owning the past and making sure people saw the past as it should be seen. For you, it was about control. It had always been."

"You looked for a cure in your genetic laboratory. I'm looking somewhere else. If I can benefit from the results of my search, what's wrong with that? I have to make money somehow to continue the search!" Sebastian strained at her straps. "Look what you did. Had an affair and gave birth to a Nephilim! Think that will cure us when the trembling starts? Huh?"

"Stop it!" Knight shouted. "Both of you."

"Nephilim?" Steel said. He looked at Knight and Renee. "What are you talking about?"

"Oh, wait until you hear the rest of the story." Sebastian said. "Go ahead, Renee. And, you, Arthur Knight, tell Mr. Steel your story. It'll make my behavior look like that of a saint."

Chapter 22

Arthur Knight straightened his bow tie, drew a deep breath and stepped through the door into the conference room. He was taking a huge chance. The eight people inside turned in unison from their places at the table to look at him. Lieutenant Commander Miller stood up slowly at the head of the table and glared at him.

"What do you want Dr. Knight. This is a private meeting."

Arthur swallowed. "I think I may have made a breakthrough, sir."

"You think?" Miller's face reddened. "Dr. Knight, you've been at this for six months and every week, you tell us you've made a startling breakthrough."

"This time, I have!" This was dangerous and he knew it. But, he had to take a chance. "I've found something in the old archival data from the 1970's. A project that addressed some of the earliest genetic research. An ex-KGB agent who developed a gene splicer. He was at the summit in Asilomar."

Miller laughed and looked around at the other officers and doctors sitting at the table. "The seventies? We didn't have the technology to dissect the human genome back then. No, you're done, Knight."

"No! Wait!" Knight hurried across the room to the table and pulled a folder from his jacket. "It was called the Anak project and it was designed to develop a super soldier."

Miller jerked the folder out of his hands and studied the contents. "Very well, you have two minutes."

Knight nodded. "Using references from the Bible, the

project determined that there were giants in the past who came from the union of the 'sons of man' and the 'daughters of men'."

"The Bible?" Miller looked around at the table. Men and women were smiling. "This isn't Sunday School, Knight."

"I know. But, listen to what they found out. There was an entire group of humans who were called the sons of Anak. This king, Anak, had sons that were over nine feet tall. These men were the ancestors of the Philistines. You remember the Philistines?"

"No. I never met one."

"Goliath! He was a Philistine. Now, the project tried to track down the exact heredity and lineage of these sons of Anak. If any living descendants of the sons of Anak could be located, the genetic material might possible produce genetically enhanced humans."

Miller slowly sank into his chair. His grin had disappeared. "And the reference to these 'sons of men'?"

"Mere allegory. The real reason they were tall and strong was their genetic makeup."

"Go on."

"The project lasted for five years. A team of what we could call today genetic archeologists went literally all over the world investigating myths of giants and they failed to locate any descendants. But, they found humans who were much higher on the normal scale for size. That's when the project was abandoned. The ex-agent disappeared without a trace along with his gene splicer. But, if we could locate some of these isolated groups of individuals today, we could catalog their genes. We have the technology they lost. We could isolate the Goliath gene."

"The Goliath gene? Where did you get that from?"

"I sort of, uh, made it up."

Miller sighed and looked around at the rest of the people at the table. "A few years back I was chasing a myth like this. That's how I met my wife. That myth was linked to this name, Anak. Very well, Knight. You have a month. I need some information by then or we close up shop. Understand?"

"Yes, sir." Knight nodded and picked up his folder. He hurried from the room and closed the door quietly behind him. The hallway was dark and empty and sweat ran down the back of his neck. What had he done? How could he possibly follow up on the myriad of leads left over from the seventies project? He would need dozens of workers and he was the only one.

"You've painted yourself into a corner, haven't you?"

Knight glanced at the far end of the corridor. A man was standing silhouetted against the window of the exit door. He slowly moved along the hall until light from an open door illuminated his ghostly face. His skin was milky white and his head was hairless. His eyes glowed a bright red. He was tall and thin and wore a maroon tuxedo over his gaunt frame.

"Excuse me? Do I know you?"

"No, you and I have never met." The man drew closer. "I'm Dr. Lucas Malson."

"Arthur Knight. I'm with the Maxsapien Project."

"Not a very subtle name. Did the information help you?"

Knight slid the folder into his jacket. "How did . . .?"

"I made sure you found it, Mr. Knight. I was a member of the task force."

Knight glanced over his shoulder at the conference room door. "You were there?"

"Yes. I knew every member of the team personally. Unfortunately, most of them are dead. But, I can help you recover the data."

"Funny you should show up at just the moment I need help."

Lucas stepped closer and placed a hand on Knight's shoulder. "My employer keeps tabs on anyone showing an interest in this line of thinking."

"Who is your employer?"

"Someone very wealthy and very well connected. He found me after the project was abandoned and encouraged me to look in a new direction. So, listen to me carefully, Arthur Knight. What the Anak Project failed to take into account when it was investigating the ancient stories of the Bible was the supernatural elements involved."

"Supernatural elements? What do you mean by that?"

Lucas smiled and exposed large, perfect teeth. "The 'sons of man' mentioned in the Bible are sometimes thought to represent creatures who were not of this world."

Knight raised an eyebrow and stepped back. Lucas' hand fell away from his shoulder. "You're not talking about aliens, are you?"

Lucas folded his arms across his chest. "Imagine if human DNA was altered by the insertion of such genetic information. Why, such a hybrid would have incredible power."

Knight's brow furrowed as he tried to comprehend what he was hearing. "Either you're nuts--"

"Or, I know just where we can find such genetic material. So, are you interested?" Lucas smiled.

Knight glanced back at the door to the conference room. He had stepped out on a limb and his future was in jeopardy. "Can I meet your employer?"

Arthur Knight fought down nausea as the small car lurched to a halt at the edge of the airfield hidden in a valley surrounded by the Andes mountains. It had taken them thirty minutes winding up and down a pothole riddled highway to reach this inhospitable spot. Dr. Malson had remained untouched by the dizzying drive. He sat rigid and unmoving in the front seat while Knight had tried to keep from getting thrown out of the back seat.

Knight popped open the door and lurched out into the fresh, cold air. He fought back nausea and marveled at the huge expanse of the Andean mountains. A shadow fell over him and he looked up into eyes the color of lilac. The man was tall and thin and something otherworldly hung about him. Hair. Not only was his scalp bare, but he had no eyebrows and no eyelashes. What was it with the baldness?

"You must be Arthur Knight."

Knight stood up from his hunched over posture. "Yes. Sorry for the appearance. Our driver was a bit wild."

"Well, I once took a night ride through the winding roads of the south island of New Zealand in a torrential downpour. Chasing down a tasty rumor about a Maori demon. As civilized as that country may be, that ride was the most harrowing I have ever encountered." The man smiled. "Sorry, you don't want to hear about my escapades. Fortunately for you, we will not be continuing this journey by road. We can reach almost to the summit by helicopter. I am Anthony Cobalt, by the way."

Knight nodded his head. "I thought you looked familiar. You just announced that rocket ship of yours. You want to be the first person to put a commercial craft in orbit, right?"

"The Aquila." Cobalt said. "Cobalt Propulsion Laboratories will succeed where others have failed."

"What does aviation have to do with my Maxsapien Project?"

"An excellent question." Cobalt smiled and gestured to the bright blue sky above them. "The stars may seem randomly placed in the heavens. Humans try and find order in the chaos and imprint stories upon those points of light. Constellations evolve over the centuries, each carrying a story. Stories are information, Mr. Knight. And, so is DNA. It is all information and I am convinced that where humanity sees chaos, there is indeed an order beneath the structure of reality." He laughed. "Actually, CPL is in search of diversification, Mr. Knight. I don't plan on putting all of my eggs in the aviation basket."

Knight glanced over his shoulder. Lucas had gotten out of the car and motioned the driver away. He was standing rock still in the middle of the tarmac of the airfield. The man gave him the willies. He looked back at Cobalt and realized that man's appearance wasn't much more reassuring. For a second he realized he was totally alone with these two strange men in the middle of nowhere somewhere in South America. This was insane.

"So, what's next?" He managed.

Chapter 23

Arthur Knight stood in the mouth of the cave and watched the sun set over the ocean. It had taken them all day to hike up the obscure mountain trail to this cave. Anthony Cobalt had personally flown them in a helicopter up the sheer side of the mountain to a flat mesa. They couldn't continue any higher due to the limitations of the weather and the thinning atmosphere. So, they had continued on foot.

Dr. Malson assured him the hike would be worth it. They would spend the night in a cave thousands of feet above the ocean. He turned away from the orange glow of the setting sun and walked into the darkness of the cave.

"Lucas, I'm really tired. I'm not used to all of this hiking and climbing. And, the altitude."

Lucas turned to stare at him. The orange light played across his pale features. "I thought you were once a marathoner."

"I was. Before my son was born. Children tend to cut into your exercise routine."

"I wouldn't know." Anthony Cobalt said from behind him. "I have never chosen to sire offspring."

Sire offspring? What kind of man talked like that? Knight glanced over his shoulder at the strange man and followed Lucas. "You have two more weeks to present your findings to the task force, Arthur. I promised you we would find one of these sons of Anak for study. And, I always deliver." He led the way deeper into the cave.

Lucas stopped at a turn in the cave and ran a flare against the rock wall. Bright green light burst from the tip and Knight whistled in amazement. The cave opened up into a huge cavern that stretched away as far as he could see. The walls were carved into doorways and windows. For three or four levels, the huge living spaces covered the walls around them. In the center of the cavern, stones were laid one on another to form walls and separate buildings.

"I present to you the Patagonian Giants." Lucas led them forward. "Legendary giants that lived thousands of years ago. Archeologists have been looking for these ruins for centuries."

"Information, Mr. Knight. The currency of the future." Cobalt pulled off a pair of calf skin gloves and smiled. "A year ago, I put a satellite in orbit with visual technology and abilities dwarfing those of the military. It was with this technology I was able to locate these caves. While archeologists dig with primitive brushes and percussion radar, I am uncovering the past from space."

Knight paused in the center of the huge cavern and listened to his footsteps echo into the shadows. "I can't believe this! You've known about this and kept it a secret?"

"Let's just say I have my reasons." Cobalt shrugged. "Now, if my research is correct there is a huge central shaft right over here." He led them into the darkness and Lucas held up the flare. A low wall surrounded by a ramp appeared in the darkness. Lucas led them up the ramp until they stood on the top of the wall. The greenish light reflected back from something in the depths of the circular hole in front of them. He tossed the flare down and it fell a short distance, bounced on a shiny surface and came to rest just inches from a dark figure buried in ice.

"Ice?" Knight said.

"The core of this mountain range is an icy river that, according to legend, was used by the Patagonian Giants to freeze their dead. Shall we climb down?" Cobalt unfurled a rope from his pack and dropped it into the pit.

Knight's mind was afire with possibilities as he followed Cobalt down into the pit while Lucas held the rope. Twenty feet below the stone wall, his feet settled on dirty ice. The air was chill and his breath streamed away in great plumes. "How does this stay frozen?"

Cobalt walked over and picked up the flare. "It is a mechanism we do not understand, my friend. Perhaps supernatural forces?"

That word reared its ugly head again. Knight shivered in the cold and thought about demons and angels and the ghosts of frozen giants. He walked over to Cobalt. The flare had melted a small puddle of water and just a foot away was the ill-defined form of a man. But, the scale was wrong. The head of the man was over a meter in diameter and the torso stretched away for almost twenty feet. Knight's mouth fell open in amazement as he slowly knelt in the dirt and ice to study the face frozen beneath him.

"This can't be."

"It is." Cobalt pulled off his backpack and set it on the ice next to the face. "Frozen in ice for thousands of years. Perfectly preserved. Do you think you can get a useful specimen?"

Knight nodded. "He seems to be almost perfectly preserved. Almost as if he could open his eyes and break free."

Cobalt tightened his grip on the rope. "That won't be happening, Mr. Knight. He is quite dead. I'm going to go back up to the cavern and set up camp while you acquire your specimens."

Knight listened to the man walk away into the distance. He glanced once over his shoulder at the sight of Lucas standing mutely on the edge of the pit. He shivered. Knight pulled a lantern out of his backpack and turned it on. He gasped as it seemed the eyes, now fully illuminated had opened. But, it was a trick of the ice.

An hour later, he had a sizable hole chipped away in the ice down to the giant's face. He had excised several samples of skin, hair, and even had pushed the thing's mouth open and swabbed the lining of his mouth. When the air hit the creature's face, it had started to deteriorate so he had to hurry. Already, the chill had penetrated his jacket and his hands and feet were numb. Knight packed away the samples in containers of dry ice and liquid nitrogen. He stored them in his backpack and pulled it over his shoulders.

"Lucas? Dr. Malson? I need the rope down here."

Silence answered him. He walked over to the wall of the pit and held up the lantern. No rope. In fact, the yawning mouth of the pit above him was pitch black. "Lucas? Are you there?"

Something cracked behind him and he jerked around. The hole with the giant's face was all he could see across the dirty ice surface. He turned back to the wall and clamped the lantern onto his jacket. He opened his mouth to shout and there was another long, cracking sound accompanied by a trembling of the ice. He whirled again and fell back against the wall of the pit as the hole with the giant's face began to enlarge. Huge cracks split the icy surface and the face of the giant shifted upward out of its hole. Chunks of ice began to fall away beneath the splintering surface. The body of the giant swiveled upward, arms breaking free of the ice. Its head tilted up and over Knight. He smelled the sudden fetid odor of accelerated decay and the head leaned down toward him.

In horror, Knight hugged the wall as more of the surface fell away into a dark void. The face slid toward him and he waited for the mouth to open and swallow him. An arm as large as a small tree trunk fell off at the armpit and landed at his feet. Ice cracked away and the entire body began to shrink into the dark pit. Something red and shiny hung in front of his face. It was a red stone about the size of a pecan suspended on a leather strip around the giant's neck. His hand closed on it and the strip disintegrated into dust. He shoved the stone into his jacket pocket as he hugged the wall and the remaining ledge of ice. He prayed the ice would hold.

A loop of rope fell in front of him. "Stick your arms through the loop." He glanced up and saw Lucas standing on the edge of the pit. He dove forward into the loop as the rest of the ice fell away into the pit. The giant's head swung toward him and caught him in the chest driving the air out of his lungs. As he faded into unconsciousness, he saw the giant's body tumble away into the dark void.

"I was out gathering wood along the trail." Lucas handed Knight his canteen. "You were lucky your torso lodged in the loop. I pulled you up as fast as I could after you passed out."

Cobalt sat beside him and sighed. "I was busy setting out our camp." He motioned to the sleeping bags spaced around the fire. "I'm afraid you fell through the cracks."

Knight sipped the water and stared into the flames of the fire burning in the center of the cavern. The pile of wood was huge and smoke drifted upward into the darkness. All around him, the once glittering icy wonder of this place now pressed down on him with a vague sense of ominous evil. "Why can't we leave?"

Lucas' red eyes gleamed in the firelight. "It's night outside. The temperature has fallen to below zero. That is why this cavern always stays frozen. That is why I have built such a huge fire."

Cobalt patted his shoulder. "We'll leave first thing in the morning and get your samples back to your laboratory."

Knight finished the water and glanced at his backpack. He had taken the red stone out of his jacket and hidden it inside. His hand slid into the backpack and he fingered the stone. His stomach growled and his hand drifted away from the stone to the other contents of his backpack. He pulled an energy bar from the other pocket and wolfed it down. "How did that giant stay frozen in the pit? Is it always below freezing in here?"

"Not in the spring and summer." Lucas stood up and heaved an arm full of wood onto the fire. Sparks spiraled up toward the distant ceiling.

Cobalt's lavender eyes reflected the firelight. "As I told you, a supernatural force kept the pit in a frozen state. It was only when you broke the surface of the ice that that force dissipated. Did you see anything else that would suggest a supernatural force at work?"

"I'm a scientist, Mr. Cobalt. I don't believe in the

supernatural." Knight hunched forward toward the fire and finished his food. He tried not to think of the red stone in his backpack.

"Then how do you explain that giant? How tall do you think he was? Ten feet? Maybe eleven?"

"Try twenty."

"Robert Pershing Wadlow."

Knight glanced at Cobalt. "What?"

"Robert Pershing Wadlow was born on February 22, 1918 to a lovely couple in Alton, Illinois. He was the oldest boy of five children but it became obvious at an early age that Robert was nothing like his siblings would be. At age 8, he was taller than his father. Why, in elementary school they had to make a special desk for him. By the time he graduated from Alton High School, he was over eight feet tall. Medicine didn't have CAT scans or MRIs back then of they would have found the tumor of his pituitary gland that produced his gigantism. He would reach the astonishing height of eight feet and eleven inches by the time he died at the young age of twenty-two. You see, his size and weight of over four hundred thirty pounds was too much for a human frame to subsist." Cobalt leaned forward and motioned to the pit. "How would a race of such beings function in the gravitational field of earth? Robert's heart and lungs just couldn't stand the strain of pumping all that blood for years. His joints broke down. Did you notice that giant you saw in the ice had gray hair? He was old. And, look at all of these rooms carved into the rock. There was a huge population of these giants in this cavern. I'd say there were at least two hundred of them."

Knight squinted into the shadowy darkness and studied the huge doors carved into the walls of the cavern. Stairways had been cut into the rock and each step was over two feet in height. "What's your point?"

"It would take more than the forces of genetics to keep these people alive, Dr. Knight. Natural forces would have been insufficient." His lavender eyes reflected the flames as his voice filled with fervor. "Supernatural forces were at work here! Those forces kept these giants alive. Those forces guarded their icy tombs. But, now that their secret has been discovered by you, there is no longer a need for protection."

Knight swallowed and peered into the rippling shadows of the cave. A race of giants flourishing here in this high altitude? How had they pulled it off? And, a pit containing ice preserved for thousands of years? He shook his head. "I can't wrap my head around that, Mr. Cobalt. So, I'm going to sleep."

He pulled off his jacket and stripped down to his tee shirt and pants. He took off his shoes and slid into his sleeping bag. He glanced once at Cobalt's glistening eyes and then pulled his backpack toward him. He reached in and palmed the red stone then snared another energy bar. He tucked them into his sleeping bag. Cobalt just glared at him with that strange grin on his face. Lucas moved for the first time since Cobalt had started his story. He smiled and revealed his impossibly white teeth.

"Sleep tight, Arthur. Don't let the bed bugs bite."

The chattering of his teeth woke him up. Knight wiped drool from his freezing lips and realized his head was out of the sleeping bag. He blinked into the shadows dancing far overhead in the vast recesses of the cavern. He turned his head slightly and his breath streamed out into the frigid air. The fire was still burning although not as large as it had been. But, what drove fear into his heart was the sight of Lucas Malson.

He was naked and sitting in a lotus position facing the fire with his back toward Knight. His nakedness seemed in defiance of the extreme cold of the chamber. But, what drove his heart into his throat was the sight of the tattoos that covered the man's back and arms. They were too numerous to count. Each tattoo was separate and small representing some object or animal. But, the eerie thing was, they were moving!

Knight blinked and tried to calm his racing heart. Was it a trick of the firelight? He watched the tattoos wiggle and squirm as if they were trapped on the man's skin. Then, he saw it. A red tattoo the shape of a drop of blood about the size of a pecan on Lucas's left shoulder blade. The tattoo wasn't moving as much as it was pulsating. And, with each pulsation, a crimson wave passed through the adjacent skin and dissipated over the other tattoos. Something vibrated in his sleeping bag. He pulled out the red stone and held it up. It was glowing with a faint red energy in perfect rhythm with the pulsing tattoo on Lucas' back. Cobalt sat up beside him and looked at the stone.

"Ah, the Bloodstone. I'm glad you found it, Mr. Knight. It is the real reason we came. I trust you will keep it safe until I need it." The man's eyes filled with malice and Knight tried to pull away but his legs and his arms were confined by the sleeping bag. He looked back at Lucas. The man's head turned completely around so that his face was directly over his tattooed back. Lucas' eyes glowed with fire in the black shadows of his face.

"Trouble sleeping, Mr. Knight?" He rasped.

Knight jerked again and opened his eyes. He was surrounded by darkness and turned his head up toward the opening in his sleeping bag. The fire had died down some and Lucas was stretched out in his sleeping bag. Cobalt was snoring in his bag just off to the side. He had been dreaming. He glanced at his watch. Four in the morning. He climbed out of the sleeping bag into the cold air and placed some more wood on the fire. He shrugged into his jacket and sat close to the flames so that he could keep an eye on Lucas. No more sleep for him this night!

Chapter 24

Renee Miller slammed the telephone down and glared out the window of the clinic at the traffic snarling downtown Dallas. Her life felt like that, twisted and chaotic. Her husband, Lieutenant Commander Miller had not been home for over two months. And now, he had just told her he was leaving for a seminar in Rome. Italy. How she would love to see Rome. She should have listened to Dr. Lawrence. He tried to warn her about Robert. She wiped away the tears and turned back to her desk and smiled at the little girl in the office chair.

"I'm sorry, Rachel. That was someone I'm not very happy with."

Rachel, a seven-year-old with the smile of Mona Lisa beamed at her. "It's OK, Mrs. Miller. My mommy and daddy do that sometimes, too."

How could the little girl know Renee had been talking with her husband? Children were way too perceptive sometimes. "Now, your tests look good and I'm going to send you back down to your mother with good news. Dr. Briner will come talk to both of you in just a few minutes."

"You did good, Mrs. Miller. The sticker didn't even hurt." Rachel pointed to the bandage on her arm.

"I pride myself on painless stickers, Rachel." Renee escorted the little girl to the hallway of their Inherited Disease clinic. She motioned to an aide and handed the chart to Rachel. "Now, you tell you mother we said 'hello' and Dr. Briner will be there in a moment."

Rachel smiled and hugged Renee around the waist. Arthur Knight leaned against the wall, biting on his finger, his eyes locked on her. Renee felt a warm tingle down her spine and unconsciously straightened her scrubs.

"Arthur, what are you doing here?"

He blinked and shoved his hands into the pockets of his lab coat but not before he touched something at his neck under his scrubs. "I needed to talk to someone. Do you have a minute?"

Renee studied his messy brown hair and his intense dark eyes. The five o'clock shadow of a beard gave him a rugged look and she tried to calm her racing heart. She was, after all, a married woman and he, after all, was a married man. She pointed to her cubicle.

"Come on in. I'm just checking the lab work on our clinic patients, but Rachel was the last one for the morning."

Arthur followed her into the office and slumped into the chair. "I don't know where to begin."

"If it's about last week--"

He glanced up at her, his face etched with confusion. "Last week?"

"When we met after work for coffee?" Renee rubbed her left hand. Arthur had reached out and placed his hand on hers last week. It had been an awkward moment and she hadn't seen him since.

"Oh, that." Arthur's face reddened. "No, it's not that. But, if I offended you, I'm sorry."

"Oh, no." Renee sat back in her chair and hid her hands. "I was going to call you and tell you something very important."

"What's that?" Arthur studied her with his intense eyes.

"I'm filing for a divorce from Robert." There. She said it our loud for the first time. In fact, she had only finalized the decision just moments ago when her husband had announced he was going to Rome without her.

"Are you sure? I mean that's pretty permanent."

A tear trickled down Renee's cheek and she looked away. "Yes. I'm tired of the way he treats me. I'm just a trophy wife to show off at the officer's club."

Arthur reached out to wipe away the tear with a tissue.

His hands rested on the far edge of her desk. He fidgeted with the moist tissue.

"I know how you feel. Claire and I have been having some hard times lately. She seems so distant. When I get home, I have to wash the clothes and cook for Josh. It's like she's become a ghost. You know what happened yesterday?" Arthur's fists clenched in anger. "I called her on her cell phone. And a man answered the phone!"

Renee's heart raced and she licked her lips. "No!"

"And then this man said I could talk to Claire as soon as she got dressed."

Renee almost smiled. This was too good to be true. "Where was she?"

"When she got on the phone she was all out of breath and said she had been trying on some clothes at Macy's. The man worked at the cashier stand and heard her phone and answered it. Now, tell me, how many men work in women's clothing? Huh?"

Renee looked down at his hands and slowly reached over to touch them. "I'm so sorry, Arthur. Do you think--?"

"She's seeing someone? Probably. I can't say that I blame her. After all, your husband keeps me down in the basement lab most of the time. And, I'm just a consultant, Renee. I was on loan from my university in Albuquerque for only three months and I've been here in Dallas almost a year! Claire and Josh go back and forth between here and home. And, I'm not part of the military! I don't belong here."

Renee spread her hands out slowly to engulf his. He relaxed as he stared off into space and his fists unclenched. He turned his gaze downward at their hands and then back up at her. He sat there in silence for a moment and then slowly pulled his hands away.

"I liked that." He mumbled.

"Me, too."

Arthur looked away and touched the center of his chest. What was he wearing around his neck? "But, I need to talk to you about something else."

Renee nodded and tried to recapture her professional demeanor. "Of course. What's on your mind?"

"I found something that may have saved my career." His gaze burned with passion. "I found something in the archives that might help with the Maxsapien Project. But, then, your husband gave me a month to dig it all up and I'm all by myself."

"Go on." Renee said.

"And then, this man shows up. Dr. Lucas Malson." Arthur leaned forward. "Renee, do you believe in God?"

Renee's brow wrinkled and she shook her head in confusion. "Yes. As a matter of fact, I do. I'm a good Lutheran." And I'm not above a little flirting with a handsome man on the verge of leaving his wife, she thought. Her face flushed with shame and she gathered her thoughts. "Why do you ask?"

"I suppose you've read the Bible, then? There was this account of the sons of Anak. They were powerful and mighty. They were giants and the nation of Israel was going to go up against them to reclaim their territory."

"Yeah, after they left Egypt. I remember that."

Arthur's hand left the thing hanging around his neck and he pointed in emphasis. "What if these men were really some kind of giant? Not just victims of gigantism. That would never work. They would have so many anomalies they wouldn't be good warriors. But, what if they had a genetic mutation that made them larger than life? What if they were truly proportioned giants with incredible power?"

Renee looked around the cubicle as if something had just entered the room. "I suppose such a thing could be possible. But, that was thousands of years ago."

Arthur reached out and took her hand. With his other hand, he reached into his tee shirt and pulled a necklace from around his neck. A large, red jewel dangled from it. He placed the jewel in her hand. "This came from a cave in South America. It was around the neck of a Patagonian giant, Renee. The man was huge. And, there was a whole civilization up there. They were real. And, I have a genetic sample!"

Renee flinched when the red jewel touched the skin of

her palm. It burned with warmth. And, it was not just the warmth of Arthur's body. "What are you suggesting?"

"Help me isolate the genetic material. If we could adapt it to the Maxsapien project, we could solve all of the problems in one blow."

Renee studied the jewel in her hand. It was heavy and hot and a nagging sensation of evil crept into the edges of her soul. This was wrong! She couldn't do it! And, then she looked up into the eyes of Arthur Knight. There was such naked ambition there; such wanton desire. Her heart melted and she chased away her reservations. She smiled. "I'm in."

Renee Miller hurried down the corridor and bumped into a tall, pale man in a lab coat. The man's face blurred past her as she rushed toward the lab. Why hadn't she heard from Arthur in the past few days? She had helped him secure laboratory space and equipment in the clinic. Her heart ached with longing as she turned the corner and paused outside the door to his lab. She slowed her breathing and opened the door.

The lab was a mess. Beakers and tubes and papers were scattered across the four lab counters in the room. She heard something break and a voice sputtering with swear words. She hurried to the last lab bench and turned the corner.

Arthur Knight was holding his bleeding hand. His hair was disheveled and he had a four-day growth of beard. His lab coat was stained with spilled coffee and his pants and shirt were wrinkled.

"Arthur?" She asked.

He glanced up at her and his teeth were clenched in pain. "Can you get me some ice?"

Renee turned and located the ice machine. She grabbed a washcloth from a cupboard and filled it with ice. She hurried over and took Arthur's hand in hers. The cut wasn't too bad. She pulled his hand over a sink and ran cold water over to wash away the splinters of glass. Then, she wrapped the icy washcloth around his palm.

"How's that? Are you okay?"

Arthur jerked his hand away and slumped onto a stool.

"No, I'm not okay! I have two more days to solve this problem or I'm out of the project, Renee. Two days! I don't suppose your husband has told you he's delaying his trip home?"

Renee's face warmed with anger. He was so abrupt and he acted like it had only been yesterday they last saw each other. "I haven't heard from my husband. It's like he fell off the face of the earth. Sort of like you did!"

Arthur looked up from his hand and his face relaxed. "I'm sorry. I'm so frustrated and I'm under so much pressure."

Renee looked away. "That's all right. I'm sure your family was glad to see you."

"Actually, I haven't talked to Claire in two weeks. She took Josh and went to California to her mother's house. I don't know why. She just said it was medically related. Her mother has breast cancer and she's not doing very well." He paused and pushed the cloth tightly into his fist. "We're not doing very well, are we?"

Renee shook her head. "I shouldn't have barged in here."

Arthur reached out with his good hand and took hers. He pulled her toward him. She looked down at her feet and the cold cloth touched her chin as he tilted her head up. She looked into his bloodshot eyes. "I need you."

"I know. I need you." She sobbed and melted into his arms. He held her as she cried until her tears were spent.

"I hope those are tears of joyous triumph." She heard a man's voice behind her. She pulled away from Arthur and wiped at her eyes as she turned. The tall, pale man leaned against the far end of the lab bench.

"Who are you?"

The man glared at her with surprisingly bright, red eyes. "Dr. Lucas Malson. I'm working with Mr. Knight. And you are?"

"Renee Miller. I'm a nurse upstairs in the genetic clinic. Arthur and I are friends, and--"

"More than friends it would seem." Lucas smiled as he walked toward them. "But, I am not opposed to anything that would help Mr. Knight complete his task."

Renee glanced at Arthur and he paled. "I'm failing, Lucas.

The Patagonian genetic material is a bust. It's human DNA. At least, superficially. Sequencing will take months even if I could bump someone in line for the sequencer. Our theory doesn't work."

Lucas raised an eyebrow and licked his lips. Renee shivered at the sight and caught a glance of the tattoo of a scorpion on the man's chest through the open throat of his shirt. "Remember what I said was responsible for the maintenance of the specimen's condition in the cavern?"

Arthur blinked and Renee realized they were trying to hide something from her. "I can leave."

"No!" Arthur said forcibly. "Dr. Malson doesn't know that you are helping me with the analysis."

Renee averted her eyes from Lucas and swallowed. "You didn't tell him?"

"No. I haven't had time." Arthur stood up and moved over to his laptop. "He doesn't know that I told you about the giant human we found in the ice caverns. He doesn't know about the sequencing attempt. What I haven't told you, Rene, is that Dr. Lucas maintains that some kind of supernatural force was at work."

Renee tried to hide her surprise. She nodded and looked directly at Lucas. "Interesting. What do you mean by supernatural?"

Lucas smiled and crossed his large hands over his chest. "Forces beyond human ability. Has Mr. Knight mentioned the Bloodstone?"

Arthur gasped and his good hand went instantly to his neck. "Bloodstone?"

"That is what it is called, Mr. Knight. I saw you take it from the giant's neck. I don't blame you for hiding it from me. All of this talk of the supernatural has you spooked." Lucas reached out and tugged at a leather strip around Arthur's neck. The stone popped into view and swung down over his white lab coat. It seemed to shimmer and glitter with light.

"The Bloodstone, Mrs. Miller. It is a part of a larger stone that, according to legend, would give great power to its user. Legend says it fell from the heavens and may be an extraterrestrial artifact." Lucas reached toward the stone and his fingers paused just short of it. For a fleeting second, his eyes filled with fear and then he retracted his hand. "Mr. Knight, if you would be so good as to clutch the stone in your injured hand."

Arthur glanced down at the bloody cloth. "What?"

"Humor me."

Arthur pulled away the bloody rag revealing a half dozen cuts in his palm. He reached up and cupped the stone in his hand.

"Now, close your fist and concentrate on the thought of healing your hand." Lucas said.

Arthur closed his hand on the stone. He closed his eyes to concentrate. "This is silly."

"Only if you think it is." Lucas said. He glanced at Renee and she shivered. His eyes were almost glowing with red fire. She looked back down at Arthur's hand and a burst of crimson light lit up his fingers. He gasped and dropped the stone allowing it to dangle around his neck. It glowed with an inner light that slowly faded.

Renee took Arthur's hand and examined it. The cuts were gone. His palm was pink and normal. "How did it do that?"

Lucas reached out a white finger and traced it across Knight's palm. "The healing process was amplified and accelerated by the Bloodstone. Do you believe me now, Mr. Knight?"

Knight glanced at Renee and excitement filled his eyes. "What do I have to do with this to prove my theory?"

Lucas smiled. "I have a retrovirus culture in the next lab. I have already inserted the Patagonian DNA into the virus. Once the virus is introduced into a human, it will reprogram the DNA in the reproductive organs of that individual. But, the process can only be initiated by the presence of the Bloodstone. You give that retrovirus to a human volunteer and have them wear the Bloodstone. Then take a sample and check it. I'll make sure you have access to a sequencer tomorrow afternoon. It's just that simple."

Chapter 25

Arthur stared out the kitchen window at the swing set in the backyard. It seemed like it had been a year since he had last played with Josh. Now, their temporary home was cold and empty; an aching loneliness painted with vacant shadows. He sipped at a cup of hot tea and his hand fell to the bloodstone. The military wasn't without volunteers. He had spent the remainder of the afternoon with a group of volunteer soldiers. Each had taken the retrovirus and each had worn the bloodstone around their neck for an hour. Lucas said it would take overnight for the changes to be introduced into the men's reproductive process. So, this time tomorrow, he might have the key to the success of the Maxsapien Project. But, the added element of the "supernatural" bothered him. He lifted the bloodstone and gazed into its dark, crimson interior. For a moment he thought he saw sparks deep within. There was a knock on his front door.

Arthur glanced through the peephole. He was excited. He was guilt ridden. Emotions warred with each other and he opened the door and Renee Miller stood in front of him. He studied her wet eyes and her wrinkled scrubs as she stood there outside his front door. He slid down a slippery slope; caution and wariness thrown to the wind. All he wanted in that moment was to hold her, to press her to him, to grip her warmth and life; to fill himself with her presence.

She leaned in toward him. His hands gripped her arms, his lips pressed into hers. They were warm and soft and the voice that screamed caution in his brain slowly faded. Without a word, she pushed him back into the living room and closed the door behind her. Her lips pressed against his face, kissing his eyes and his cheeks. Tears poured down his face and she stripped away his shirt.

Something hot stung his chest and he glanced down at the bloodstone. It glowed red with energy. And then, the voice came. Quietly at first, it spoke in the back of his mind.

"Give in. Take this. You deserve this. Give me control and we will make this happen. You want this. You want this so badly. Show Claire what she gave up when she threw you away."

Arthur listened to the voice and its warmth and energy wrapped itself around his face, enclosed his nostrils and mouth with moist air. He inhaled it and the power filled him, burning the bloodstone at his chest, making him giddy and joyous with happiness. He turned his face toward the ceiling and laughed out loud as he pulled Renee to him.

"What have we done?" Renee sat on the side of the bed. Remorse and guilt filled her mind, chasing away the passion and joy of the past hour. His hand rested on her back.

"We did what we should have done long ago." Arthur said. His voice was peculiar, raspy. She turned and studied him. The bloodstone glowed gently on his chest.

"Arthur, we're both married. You have a child!"

"I didn't hear you protesting about that an hour ago." He sat up. "Come on. We knew it would happen. We knew it was inevitable. When Claire comes home, I'll file for divorce. I'm sure she's seeing another man. Your husband has all but abandoned you. It shouldn't be hard for us to end these marriages. They don't work!"

Renee wanted to believe him. She wanted to walk away from the pain of her past, but something felt so wrong. So very wrong. She shook her head. "No. This is wrong. We shouldn't have done this."

Arthur laughed and slid across the bed toward her and looked up into her face. "Come on. We were great together. We should have met each other years ago. What a lot of wasted time we've had. And, let's not waste anymore tonight."

Renee pulled away from his grasp and glared at the bloodstone. "It's that stone around your neck. It's been glowing ever since I got here and you're different."

Arthur frowned. "Different? What are you talking about! I'm just free from the past. Rene, I don't care anymore about giving in to your husband or the project or my manipulative wife. I'm free and so are you."

Renee watched the stone glow more brightly. "If you are free, then take it off."

"What?"

"The bloodstone. Take if off and hand it to me."

Arthur's face froze and his hand drifted to the stone. He looked away and something foreign crossed his face. "I don't want to."

Fear grew within her. "I did some research of my own. I read that passage in the Bible about the giants. They were called the Nephilim. It says that women were seduced by demons, Arthur. And, the children were called the Nephilim. What if this stone puts you under some kind of spell?"

"Oh, get off it. A demon spell? Come on, Renee. We're both scientists." Arthur said.

"Then if you don't believe me, take off the stone. Now."

She watched his hand drift up to the bloodstone. He glanced down at it. "You're being ridiculous."

Renee shook her head as tears began to stream down her face. "Please, Arthur. If you care about me at all, take off the stone! Now!"

Arthur just glared at her and smiled a lopsided grin. "No."

Renee's hands came up to her mouth. She slowly backed toward the door to the bedroom. "Arthur, I'm going. Now. Before you do something you regret."

Arthur stood up slowly and extended his hands toward her. "Come on, Renee. It's just a piece of rock. You're overreacting."

Renee stifled a sob and hurried through the doorway and back into the living room. She tripped over Josh's skateboard and fell sideways onto the coffee table. Pain lanced up her elbow as she hit the edge and rolled over onto the floor. She gasped with pain and looked up into the half crazed eyes of Arthur Knight.

"You're not going anywhere, Renee. We're spending the night together." He stood over her and the bloodstone swung like a pendulum between them.

She crawfished backwards around the corner of the couch and pulled herself up by the door handle. Her hand was numb from the blow to her elbow and she fumbled to twist the knob. The door flew open onto the night and she stumbled down the stairs and ran into Lucas Malson. He caught her with his huge, pale hands and she looked up into his red eyes.

"Leave me alone!" She screamed as she tore out of his grasp and ran to her car. She fumbled in her scrub pants for the keys and sped away into the night leaving Arthur at the mercy of the devil himself.

Arthur settled onto the couch as Lucas kicked the skateboard out of his way. He wore a long, black overcoat and a black shirt and pants underneath.

"So, how was your evening?"

Arthur held onto the Bloodstone as he tried to chase confusion from his mind. He had wanted to hurt Renee. Why? It was as if something had taken over part of his mind. "Good. I think. What is happening to me?"

"The Bloodstone is giving you everything you've ever wanted, Arthur." Lucas sat on the edge of the coffee table and leaned into Arthur. His crimson eyes were filled with a passionate fire. "Everything, Arthur. And, you can keep it. All you have to do is invite it in to stay. Forever."

Arthur blinked as his mind rolled with conflicting thoughts. "Is it already here?"

"Yes. Doesn't it feel good?"

Arthur tried to stand up and felt weak and dizzy. "I don't know. I'm confused." He looked down at the stone hanging around his neck. "I can be like I was with her?"

"Yes." Lucas slid out of his coat and unbuttoned his shirt.

Dozens of tattoos moved on the white skin of his chest. He turned around and revealed his back. The tattoo of the red stone Arthur had seen in the cave was glowing with red light. "You can become a permanent member of my collection. You can have power unlike anything you have ever known."

Arthur drew a deep breath and the telephone rang. Lucas's head swiveled and looked at the phone on the coffee table as if it were a poisonous snake. "Don't answer that."

Arthur looked down at the blinking light. He really should answer it. It might be, who? He tried to clear his mind. Who was calling? What was her name? The phone rang again and went to the voicemail. He heard his own voice telling the caller to leave a message. That voice seemed so foreign now. It seemed to come from another time and place. Then, a tiny voice filled the quiet of the room.

"Daddy? Are you there? It's me, Josh. Mommy told me not to call. She said it's later there than here but I wanted to talk to you. Something is wrong and I wanted to talk to you. Are you OK, Daddy? Please answer me."

Arthur dropped the stone from his hand and it fell against his bare chest. Something rustled in the background on the answering machine and he heard Claire's voice. "Josh, I told you Daddy is probably asleep. Give me the phone. Arthur, sorry we called so late. I'm coming home tomorrow and we need to talk. I have something very important to tell you. Love you. Bye."

The voice died and the dial tone took over. Arthur glared at Lucas. "What have you done to me?"

"Just what you wanted. Just what your kind always wants. Fame. Glory. Riches. Sex. Now, Arthur, before you go and take off the Bloodstone and undo everything we've started, sleep on it. Keep it until tomorrow afternoon. We'll have the test results on the volunteers by then. Just a few more hours before you decide."

Arthur's mind clouded and fatigue come over him. He fell back onto the couch and blinked. The Bloodstone burned into his chest as Lucas leaned over him even as he had leaned over Renee just moments before.

"Sleep, Arthur Knight. Just go to sleep."

Renee walked into the cafeteria at two the next afternoon. Arthur was waiting for her, his mind still clouded and confused. He vaguely remembered the night before and he wasn't sure just what had happened between the two of them. He had paced impatiently in his lab all morning waiting for the results promised by Dr. Malson. Finally, he had called Renee and asked her to meet him for coffee.

She had her hair pulled up on the back of her head. She wore a light green pair of scrubs and her white coat. She stopped when she saw him and hesitated. He stood up and crossed to her.

"Rene, we need to talk about last night."

Her eyes filled with tears and she glanced down at his chest. He knew she was looking for the Bloodstone but it was safely tucked away beneath his shirt and tie.

"What happened to you?" She whispered.

He pointed to a table. "Please. Sit down."

She followed him and shrugged away the touch of his hand when he tried to help her into the chair. "What happened with Lucas last night?"

Arthur sat slowly and rubbed his forehead in confusion. "Lucas was at my house last night?" He sifted through his memories. Yes, the man had come by the house. Right after Renee had left. "I don't remember. Listen, Rene, I'm very confused. What happened between us last night?"

Renee slowly lifted her gaze from her hands. "You don't remember? Arthur, we ended up in your bed!"

"So, it wasn't just a dream."

Renee leaned forward and her face twisted in anger. "No, it was real. Very, very real. And, it was good. And, it was right. Until afterwards when you turned into some kind of monster. You don't remember that?"

"I can't remember much. It's all jumbled up. Like--"

"Someone else was in control?" Renee finished for him. "Like maybe a demon?"

Arthur blanched and a cold wave of panic flowed over him. "I don't believe in those things."

Renee reached forward and flipped his tie out of the way and with one quick grab pulled the Bloodstone out of his shirt. Arthur panicked and grabbed her hand. He couldn't let her have the stone. Not the Bloodstone. It was his. It was his power. It was his reason for living.

"Look at yourself, Arthur. You're grasping for the thing like it's some kind of lifeline. You're obsessed with it. You're possessed by it. You've got to take it off and give it up. If you don't, you'll lose yourself."

Arthur pulled her hand off the stone and the comfortable heat of it nestled in the palm of his hand. "I don't know what you're talking about."

Renee placed a palm on Arthur's chest. "Your heart is right here. I can feel it pounding. It belongs to you. You can give it to anyone you want. You once gave it to Claire. You wanted to give it to me. But, instead you've given it to Lucas Malson and this cursed Bloodstone. Take it off, Arthur. Take it off, now."

"Take what off?"

Arthur whirled. Claire stood in the middle of the cafeteria. He shoved the Bloodstone back under his tie and heard Renee gasp. He stood up.

"Claire?"

She was pale and seemed thinner than the last time he had seen her. Her strawberry blonde hair was cut short around her face. She walked toward him and her gate was slightly unsteady. He hurried to her and hugged her. He could feel the bones through her skin. What was going on?

Claire pushed him away and glanced over his shoulder. "What does this woman want you to take off?"

Arthur blanched and turned to see Renee rise from her table. Renee walked over and held out a hand. "You must be Mrs. Knight. I'm Renee Miller. Arthur works for my husband, Lieutenant Commander Miller."

Claire took her hand and nodded. "Okay. Now, can you answer my question?"

"I want Arthur to take the restraints off on his project personnel. I've tried to tell him that my husband will give him more people to help him with his project, but he's stubborn. He doesn't want to fool with training new assistants. I have two lab techs up in the Inherited Disease clinic who would love to work with Arthur on his project." Renee smiled but Arthur knew it was forced.

Claire nodded and looked warily at Arthur. "I see. Well, it's nice to meet you. If you don't mind I need to speak to my husband in private."

Renee stepped aside and cast one last desperate look at Arthur. He ignored her and settled at the table with Claire. Claire put her purse on the table and kept her eyes downcast. She drew a deep, shuddering breath.

"I--" Arthur said at the same time she said, "Arthur--"

They laughed nervously. Arthur reached out and touched her hand. His heart was pounding and the confusion began to cloud his thinking. "You go first."

Claire sighed. "I haven't been a very good wife lately, Arthur. I'm so, so sorry. Something has happened. At first, I thought I could control it. I thought I could deal with it and leave you out of things. You must understand," she looked up into his eyes. "I only wanted to spare you the pain and suffering. I was only thinking of you."

Arthur's face warmed. He opened his mouth to speak and shut it again. Claire's grip tightened on his hand.

"I didn't go to California just to see my dying mother, Arthur."

"I know." He whispered.

Claire pulled her hand away. "You know? Did Josh tell you?"

Arthur glared at her. "Oh, he knows and I don't? What did you do? Introduce him to the man who was to become his next father?"

Claire's face reddened and her hand came up to her mouth. "Is that what you think?"

"Why shouldn't I? I called you week before last. A man answered. And I heard him tell you to get your clothes on.

And, you expected me to believe you were at the department store trying on clothes?"

Claire's eyes filled with tears. "Oh, Arthur, of course I shouldn't expect you to believe me. When we choose to hide things; when we choose to lie to the people that we love, this is what happens. There isn't another man. That was my doctor who answered the phone. I went to California for a second opinion."

Arthur's mouth fell open. "What?"

Claire reached out again and took his hand. "Arthur, I have multiple sclerosis."

All kinds of conflicting thoughts flew through his mind. He saw himself in the arms of Renee Miller. He saw himself prostrate before Lucas Malson. Something foreign and ugly reared its head in the back of his mind reminding him of his indiscretions. Guilt seared his soul as the prospect of losing this woman he had loved more than life. He reached under his tie and pulled the Bloodstone roughly away from his neck. The leather strip broke and he held the stone in front of him.

"What's that?" Claire leaned back.

"My own personal demon, Claire. I have made such a fool of myself. I have made so many mistakes. And, I can't blame this little stone. I have only myself to blame."

A shadow fell over him and he looked up. Dr. Lucas Malson towered over him. His face was more pale than usual and he wore the long, back coat over his dark shirt and pants. "I see you've made your decision, Mr. Knight."

Claire looked up at the man. "So, this is your new associate?"

"Yes." Arthur stood up. "I have made my decision. My wife has brought sense back to me."

Lucas smiled and his perfectly chiseled teeth gleamed in the fluorescent light of the cafeteria. He ran a bony hand over his bare head. "Good. It seems the experiment was a success. We have enough genetic material for twelve offspring, Mr. Knight. Your plan has worked."

Arthur glanced down at the stone glowing redly in his hand. It beckoned to him. It called to him. A voice in his head begged him to put the stone back around his neck. A tear fell from Claire's face and landed on the stone with a warm splash. He heard the tear sizzle and the red light died out from the stone. He looked into the eyes of the woman he loved. Something snapped in the back of his head and he exhaled. A long, drawn out breath came out of him and he grimaced at the foul taste. His mind was clear and sharp again. He blinked and looked down at the stone. It was dark. He felt a sharp pain in his other hand. He held it up and there were the cuts from the broken flask. Blood ran from his palm and dripped onto the table. He laughed.

"My hand is bleeding. Claire, my hand is bleeding!"

She shook her head. "I don't understand."

Arthur looked up at Lucas. "Take your project and run it without me. I'm going back to the university and back to my family."

Lucas frowned. "You can keep the stone, Mr. Knight. It had no special power. It was the thing you allowed into you that was the source of that power. It seems you have rejected it." Lucas sighed. "Well, you can't win them all. But, I would be careful, Mr. Knight. Your soul is running on empty and sooner or later you will fill it up again. If you fill it with one of my associates, that stone will come back to life."

Lucas turned and walked away through the crowd in the cafeteria. Arthur looked around at the people talking and visiting over coffee and pastries. How many of them had turned their backs on their spouses because of difficulties? How many of them were having an innocent cup of coffee with a sympathetic ear? Tears filled his eyes and he shoved the stone down in his pocket. He looked up at Claire.

"I have something to tell you and I hope you can forgive me. I made a very big mistake. And, before this night is over, I want to fill myself up with something more powerful than one of Lucas's demons. Can you help me?"

Claire smiled and her tears glittered in the light. "It's what I'll live for."

Renee looked at the blue strip. Tears sprang to her eyes. Arthur had been gone for four weeks now. He had returned with his family to Albuquerque, New Mexico and to his university teaching position. Her husband had been back from his travels for the past three weeks. But, if what this little blue strip said was true, the baby wasn't her husband's.

She tossed the strip in the trash and stood up to look at herself in the mirror. Her face was streaked from crying and her eyes were red with fatigue. What was she going to do? Arthur had been wearing the Bloodstone. He had been acting strangely. What if he was demon possessed? She ran out of the bathroom and out the back door of her small home into the backyard. It was dark and the stars were soaring overhead and the night was hot and humid. She couldn't breathe and she gasped for air as the implications dawned. She put a small hand on her abdomen and settled slowly into the grass. Her sobs echoed around the back yard.

"Renee?" She heard him behind her. "What are you doing out here?"

She looked up over her shoulder at her husband. "I needed some air."

He stared down at her from a great height, his face hidden in shadow. "It's hot out here. Let's go back inside and talk about this." He held up the pregnancy test.

She rose to her feet and jerked it out of his hand. "What are you doing with this?"

"I saw it in the trash in the bathroom. It's positive." His voice was low and his eyes were hidden in deep, dark sockets. "How far along are you?"

Renee bit down on her upper lip and thought furiously for an answer. "When did you come home? Three weeks ago? I'd say I'm probably about three weeks."

"That's too early for this test."

So, that was it, she thought. She looked at the test strip. "It can be positive as early as two weeks."

"Did you miss?"

She glared at him. "What is this? An interrogation? Am I one of your soldiers?"

"Go inside. Now." He turned and pointed.

"Or what?"

He turned back to her. "I want you to go inside."

"And, I want a divorce." She said. He was silent. A hand went up to his face and she cringed.

"I already know that. I found the papers last week. You filed them after I came home."

"That was before I knew about this." She held up the strip. "But, this doesn't change anything. I can raise our child by myself. I can give it the love and attention that you give to your military. We don't need you."

He reached out and grabbed her arm in a vise like grip. "I won't let this happen."

Renee winced and twisted her arm in his grip until he let go. "Good! Now I'll have bruises to show my lawyer. Get out of my way and get out of my house." Renee stormed past him praying that he wouldn't come after her. She prayed that he wouldn't retaliate. And, as much as she feared it, she prayed that the life that was in her was not his child.

Arthur Knight sprinted the last one hundred feet up the mountain path and ended his run just short of the ski run. It was a beautiful May day along the northern side of Sandia Peak. The ski season had been over for two months and the snow still clung to the shadows of the fir trees. But, the temperature was in the low seventies and there was no humidity. He had parked his car at the base of the ski slopes and run down and back up the highway. The pace was grueling, but he luxuriated in the pounding of his heart and the regular passage of air in and out of his lungs. Life was so good in spite of Claire's problems. She had been in remission for months now. Josh was growing like a weed and the world was a wonderful place. He had become a Christian after the Bloodstone incident and had put all of that behind him.

As he walked up to his car he noticed another vehicle parked nearby. This wasn't unusual. Albuquerque was famous for its triathletes and to run the Sandia Peak highway was almost a requirement. The car door opened and his past stepped into his life.

"Renee?" He said.

Renee Miller closed the car door and looked at him. "Hello, Arthur."

Arthur's heart raced and a dull ache filled his stomach. It was the vestiges of guilt and shame he had worked through. Claire had long since forgiven him. He thought he had put his painful past behind him. "What are you doing here?"

She wore a long, loose cotton dress with sunflowers on it. Her hair was cut short in a bowl cut. "You look good, Arthur."

"You look good yourself." He stopped just two feet from her. He was aware of the sweat running down his bare chest.

"I see you still have the scar."

Arthur looked down at the red scar nestled between his breasts. The Bloodstone had burned it there. "It won't go away."

"I know. I have a problem and I need your help. That's why I'm here."

Arthur nodded. "Okay."

She reached into a large purse hanging from her shoulder and took out a photo and handed it to him. It showed a toddler in a diaper with long, curly blonde hair. The face was beautiful and the eyes a deep, jade green. "Who is this?"

"My daughter." Renee said.

"She's beautiful. Are you still married to--"

"No. I divorced him right after you left."

An uneasy feeling surfaced as he studied the picture. "Who?"

"Is the father?" Renee finished. "I believe you are."

Knight stepped back and drew a deep breath. He studied the photograph. "Is this about money?"

Renee stared at him and tears filled her eyes. She looked

away. "How can you say that? You know me better than that." She snatched the picture from his hand and wiped away a tear. "I got everything from Miller. He had that startup commuter air service. It's now mine. He had a huge inheritance from his father. I have half of it. I'm thinking about starting my own business in genetic counseling and engineering. I'm going for my doctorate." She looked back at him and her jaw was set firmly. "I don't need your money. I don't want to have anything to do with you. I need to find Dr. Lucas Malson."

"Lucas? Why?"

"Those diapers are adult size."

"What?"

"Vega is two years old. She's four and a half feet tall. She weighs ninety pounds."

"What?"

"She's a Nephilim, Arthur. And, I've traced down twelve more children just like her. They're all Nephilim. Scattered around the country. Hidden from the public. All of them Lucas' Nephilim. They are children of the Bloodstone, Arthur."

Arthur blinked and looked around at a day that had started out so promising. "This can't be."

"What did Lucas do with your research?"

Arthur tried to recall that last day in the cafeteria. What had Lucas said? "He said the experiment had been a success. He said he had enough material for twelve children."

Renee nodded and her eyes were flinty with anger. "Well, there are thirteen. And, I want her to be normal. I want a cure, Arthur. And, you are going to help me to find it. I'm starting this organization. It'll have to be secret to hide them from the government. The children will all have to be contacted and taken into hiding. I'll use my husband's money against him. And, you will find them for me. You now have a new job, Arthur. You can tell Claire whatever you want but this is your doing. You brought this on me and on the world. And, you are going to help me fix it. We're going to find these children and move them to a secret place. We'll move them from town to town if we have to. I purchased Ingenetics Institute and I'm refitting the building for research. We will find a cure before they get too large. Do you understand?"

Arthur watched his wonderful, joy filled world crumble

around him. Who knew that one small, almost insignificant bad decision could ruin so many lives?

Chapter 26

"The sons of God in Genesis 6 were humans invaded and possessed by fallen angels in such a way as to alter the genes transmitted via intercourse. In this way they produced offspring with the physical characteristics associated with the Nephilim."

Dr. Hugh Ross, "Navigating Genesis"

Dr. Liz Washington realized her mistake the moment she stepped out of her car. The sun set along the western shore of Cross Lake and painted the sky in streaks of crimson and yellow. She wore a pair of old, beat up jeans and a worn, comfortable scrub top from her days in the laboratory. The back door to the lake house had been kicked in. Broken glass from the half window littered the back door landing.

"You're no spring chicken, Liz." She whispered to herself. "Call 911." She swallowed. And, then she heard the scream. It was muffled and far away. Her heart raced and she ran into the kitchen. The muffled scream came from the stairs leading down into the basement.

Liz paused at the top of the stairs and pulled out her cell phone. She dialed 911 and waited for the operator. "This is Dr. Liz Washington." She gave them the address and waited. "Yes, there is someone screaming in the basement. No, I won't go down there until the police arrive. Yes, I'll stay on the line."

Another scream echoed up the stairs. Liz jerked and

dropped her phone. It clattered down the stairs into darkness. She heard a dial tone. The phone had hung up on the operator.

"Okay, Liz, let's go down slowly and quietly and get the phone." She went over to the far right hand side of the stairs and grabbed the handrail. Slowly, step by step she eased down the stairs. She stopped and pressed her hand against her mouth. In the middle of the stairs right beside her was a streak of blood. It was still sticky and shiny. Not a lot of it, but it was blood.

She continued down the stairs until she reached the bottom. The basement looked like a tornado had torn through the walls and the crates of Cephas' artifacts. Splintered wood and packing material covered everything. The scream came again and she picked up her cell phone and put it on flashlight setting.

The eerie pale, blue light played over the chaos around her. She stepped gingerly over the broken remains of Cephas' desk. The wall to her right was covered with something. She stepped closer. Dirt, rock, and debris stuck to the wall spreading out from something that moved and jerked in the center of the mass. A sticky substance oozed from the edge of the plastered mass. She reached out and touched it. Blood! Fresh, red blood!

The scream came again and Liz pointed her phone at the center of the plastered mass of trash stuck to the wall. Something bulbous moved at the center of the mass. She stepped closer. The spherical object was stuck down and matted to the wall by fibers of wood and dirt. Beneath the debris some kind of glistening material covered the sphere. It lurched again and she saw the material retract just before the scream. The area of retraction bulged out with the scream. The material was covering a mouth!

She tried to step away and something at the edge of the debris ripped loose and a hand grabbed her ankle. She tried to pull away but the grip was unrelenting.

"Help!" She screamed. "Someone help me!"

"Liiiiizzzzz?" The bulbous thing screamed. "Theeeooooo! Hellllllppppp!"

Liz froze. "Theo? Oh, my! Theo!"

She dropped the phone and tore at the fibers and the twigs until her fingernails dug into the rubber like membrane. It vibrated and moved with a life of its own and she realized it was supernatural in origin. "Listen here, you spawn of hell, I am Liz Washington a daughter of Eve and I command you to release my friend in the name of my Lord and Savior Jesus Christ!" She shouted.

The membrane glowed and tightened becoming solid until it cracked into a million zigzag lines. The sphere jerked and the material fell away. Theo looked up at her with blood shot eyes. "Mama Liz, you the sweetest sight for sore eyes I ever saw."

"We'll be landing in ten minutes." A voice said over the intercom.

Knight glanced at Steel. "Somehow, Miller had gotten his hands on the children. So, I started tracking down Miller's actions. I located all of the children and we placed them in a secret location. I did this until Miller figured out who I was. That is why I had to fake my death. There's no telling what Miller would have done to Claire and Josh to find me. After I faked my death, I continued to help behind the scenes. Renee didn't even know I was still alive. I posed as one of my operatives who took up the cause after my death."

"You were a fool, Arthur. Why did you ever trust Lucas in the first place?" Steel said.

"Power. Fame. Money. They're all the currency of a fool. We can't all be as perfectly focused as you, Steel. You've never had to work with the likes of Miller." Knight asked.

If they only knew, thought Steel. The plane lurched as it touched down on the tarmac. What had he been willing to do to find his father? The thought triggered a memory. He was somewhere else.

Hot Steel studied the GPS readout on his cell phone. A text had directed him to rendezvous with the second group after they had secured the second of the three targets. The third target was located in a huge drainage culvert beneath the ski lift. The rendezvous point was behind a large boulder overlooking a cross trail. The culvert emptied directly on the other side of the cross trail from the boulder. The third target's homing device located it at the culvert.

He eased through the low-lying undergrowth beneath the silent ski lift. Chairs swung in the breeze and cast dizzying shadows in the moonlight. He arrived at the boulder and no one was there. The boulder was over ten feet tall and the cross trail and culvert were on the down slope side of it. He donned his night vision goggles and surveyed the surrounding underbrush and trees. They were empty. Where was the second team? Had they been wiped out like his first team? What power did these children have?

He heard a high-pitched voice down slope. Silently, he climbed the boulder and inched over the top. Beneath him, the cross trail was covered with ruts and short grass. In the winter, it would be an easy trail for skiers linking different slopes on the mountain. In the summer, it was just another dirt road. A solar powered night light on the ski lift was on half power and cast a dim circle of light on the cross trail. A little girl stood in the center of the circle of light. She was large for a child, at least four and half feet tall. Her head was turned upward toward him and her huge eyes gleamed in the light. She wore a red jumpsuit like the boy's.

"There you are. I've been waiting." She said. Her voice was small and musical and oddly strange for her size. She seemed like a perfectly proportioned huge doll of a child. She turned her head to the right. "Be quiet! I told you I wasn't talking to you." She looked back at Hot Steel. "The invisible monsters don't want you to be here. But, I don't care. You're a good man and they can't hurt you. They tried to make me hurt you but I told them 'no'. No, no, no!" She shook her finger at the air to her right and then repeated the same phrase to her left.

"Now, you can come down here and talk to me. Nobody else is coming."

Hot Steel glanced behind him and the trail was empty

beneath the ski lift. He skittered down the front side of the boulder and stepped out into the circle of light.

"You are safe in the light. They don't like the light. That's why they hide in our closets. It's always dark." The girl whispered.

Hot Steel pushed his goggles up onto his forehead. "Who are you?"

"My name is Vega Miller." The little girl said. "I'm four years, two months, and three days old." She reached into a pocket on the side of her jumpsuit and pulled out a picture. She handed it to him. "That's my mother. Have you seen her?"

The picture showed a woman in surgical scrubs holding a large baby. He shook his head. "I'm sorry, honey. I've never met your mother." He handed the picture back to her. "How did you get here?"

Vega pocketed the picture. "That mean man with the pipe picked three of us and told us to play hide and seek." A tear leaked from her right eye and she turned tear filled eyes on him. "But, they started to chase us with bad soldiers and guns! Timmy got captured and so did Altair."

"How do you know that?" He asked.

Vega glanced to her right. "The invisible monsters told me. They said that you shot Timmy but I understand it was to keep him from killing more soldiers. The monsters didn't tell Timmy they would kill people. We thought it was a game!"

Hot Steel glanced to the right and left. Nothing stood on the road. But, he had an uneasy sensation of something just outside the perimeter of light. He shook his head to clear his thinking. He had not infiltrated this group to rescue children. He had come for one reason.

"This man with a pipe, did he have a name?" Hot Steel asked.

"Everyone calls him the Captain." She said. "And, he has real pretty blue eyes. But, he is the meanest and baddest man I have ever met!"

He had found his father. Now, if he could somehow use this girl to get to his father, he could take the man hostage and get some answers. He had so many questions and the presence of this unusually precocious child just raised even more. Vega tugged at his sleeve.

"Will you take me to my Mommy?"

Hot Steel hesitated. "I don't know where she is."

Vega glanced to her right. "He's coming! A bad man but he doesn't want to hurt me. The monsters say he is looking for me but he doesn't want to hurt me." She looked back at him. "They say he's hoping none of the soldiers are here yet. Maybe you should hide until I talk to him."

Hot Steel shook his head. "I'm not going to--" And, then, he realized the approaching man might be his father. "Good idea." He scrambled quietly up the boulder and slid down on the far side to a ledge on which he could still peer down at the cross trail. A hand slid over his mouth and lips pressed against his ear.

"Don't make a sound. Call sign?"

The hand relaxed and he instantly realized the other soldiers had arrived. He knew there was a knife ready to sever his brainstem if he didn't answer correctly. "Hot Steel." He whispered.

The hand withdrew and he heard the metallic sound of a knife being sheathed. "Stoneheart."

Hot Steel glanced over his shoulder. The man was squatting on the narrow ledge. His face was obscured by dark pigment and his eyes were covered up by a pair of night goggles. "Where have you been?"

"The second target killed everyone but me." The man was breathing rapidly. "I knew that girl had some kind of power but I've never seen anything like this. Man, what a rush."

"How'd you get away?"

"I was lucky, I guess. Got all five shots off before she could turn on me. Miller showed up and told me to proceed to the next rendezvous. Looks like your group had as much success as mine."

"We got the kid." Hot Steel mumbled. He looked back up at the top of the boulder. "The last target is right over the boulder."

"I heard it all, Hot Steel. 'You're a good man.' 'My

monsters won't hurt you.' You're lucky she didn't cut off the top of your head and eat your brain. Now, we're going to listen in on this little conversation. And, if she comes after me, you have to finish this."

Hot Steel nodded and they quietly eased up to the top of the boulder. Something about the man was eerily familiar but his naked bravado was disturbing. But such was the nature of Miller's mercenaries. You had to be disturbed to work with the man.

Major Miller stood in the circle of light with Vega. She had backed to far edge of the circle keeping a distance between them.

"Vega, we have to talk before they get here."

"How do you know me?" She whimpered.

"I'm your father, honey. Don't you remember?" There was strange pleading tone in his voice.

"What is this crap?" Stoneheart whispered. "He's got a perfect shot."

Miller glanced in their direction and Hot Steel and Stoneheart ducked out of view. They eased back up and Hot Steel motioned with a finger across his throat for Stoneheart to be quiet.

"You are not my father!" Vega shouted.

"I was there when you were born." Miller reached out a hand.

"Mommy told me about you. She told me you were a mean man and you were a bad man. She told me you weren't my Daddy. Somebody else is my Daddy." Vega pouted.

Miller seemed to shrink, drawing back his offered hand. "I thought as much. But, that doesn't change the fact I am here to protect you. The soldiers will be here any minute and they will take you back to the Captain. Once he gets you I can't protect you. Understand? I can tell him we couldn't find you. I can tell him you won the game of Hide and Seek."

"No!" Vega stomped her feet. Then she whirled and looked out into the darkness. "No, you will not hurt him! You do not do bad to somebody because of me. I'm a good girl and you are very, very bad. Go away! Now!"

Miller stepped closer and Vega turned quickly. "They

want to kill you, but I won't let them. But, I can't stop them forever. Go away! I'd rather be taken by the Captain than call you my Daddy. Go away! Just leave me alone!"

Miller stood up and sighed. "Very well. But, understand that from now on I can no longer protect you." He turned and without hesitation walked into the darkness.

"How touching." Stoneheart whispered in Steel's ear. His anger surged at the familiarity in the voice and he whirled.

"Do I know you?"

Stoneheart's camouflaged face was inches from his own. He smiled.

"Now, listen carefully Mr. Hot Steel. We don't have the final coordinates yet. Miller will send them to us shortly. By then, I will be where I need to be. I'm following Miller, understand? I have some unfinished business with a certain pipe smoking captain. So, you have to deliver little missy down there to her final resting place. Once you have her, signal Miller and he'll give you the final coordinates. And, when you get there remember I have my own agenda, man of steel. And, it doesn't involve you. Just the Captain."

Hot Steel opened his mouth to protest and he heard the girl's voice. "Help me, Mister. I know you're there."

Stoneheart grinned and dropped off the ledge into darkness. Hot Steel swore under his breath. Miller would lead him to the Captain. And, this other soldier was after the man, too? That shouldn't have surprised him. His father had many enemies. But, the girl needed help. He had not come here to rescue a child, he reminded himself. He shook his head and hopped over the top of the boulder. He came to rest in the circle of light. The little girl was waiting for him, her arms crossed over her chest.

"The invisible monsters are gone. They don't like me. And, they don't like you. But, they really like that other man with you." She blinked at him with her huge eyes.

"Listen, Vega, I didn't come here to rescue you. They want me to stun you and take you back to the Captain."

She shook her head. "You're going to take me back and make sure my Mommy finds me. That is what the invisible good person told me. He said that you were going to get me to Mrs. Donnelly and my mother."

"What invisible good person?"

Vega shrugged. "I can see them all. Good or bad. There's a good one right there." She pointed past him and he turned around. The trail was empty. The hair stood on the back of his neck and Vega took his hand. She nestled up beside him.

"Okay. I'll tell him." She said quietly. Steel looked into the girls' eyes.

"Who are you talking to?"

"Your guardian angel. He wants me to tell you something." Vega looked back at the empty space. "Oh yeah, I'll tell him that, too."

Steel glanced at the empty air. "This is crazy."

"He says that things are only going to get worse and that one day you will remember this night. When you do, you will understand what your mission is." Vega smiled.

"My mission?" He looked back at the empty air. "What mission?"

"He said you'll know it when the time comes. He said that much you won't be able to forget. So, let's go."

Hot Steel opened his mouth to speak and the cell phone beeped. He slid it out of his pocket and looked at the map. Now, he knew where to find Miller and the Captain. But, the little girl needed her mother. She had faith in him, of all people, to reunite her with her parent. Could he do this? What about the Captain? He glanced back at the empty air. Guardian angels? Yeah, right! He sighed and reached out his hand. "Come on. It'll be a long walk."

"You're not going to stun me?" The little girl's huge hand almost dwarfed his.

"What, and carry you? You must weigh close to one hundred pounds. No, Vega, you're going to walk. And, you're going to tell me about this invisible friend of yours."

They stepped out of the light into darkness.

"You coming?"

Steel shook his head and glanced up at Sebastian. The others had left the cargo plane and were standing on the tarmac.

"Yeah, just reminiscing."

He unbuckled from his seat and followed Sebastian down the ramp. He had met Vega in the past and he had actually helped capture these children. He savored the feeling of hatred he felt for his father. He watched Knight standing off to the side away from Renee and Cassie. Was he afraid Josh would hate him for faking his death? If so, then Josh might hate his real father as much as Vega had hated Miller in Steel's flashback. As much as he hated his own father, the Captain.

But, here was one thing he was sure of. Vega's "invisible friends" were a force he was familiar with. Fortunately, she had conquered her demons while he still fought with his. But the memory that seared his soul; that burned into his heart was the statement made by the "guardian angel". He would know his mission. Now. He sighed and tried to relax. He had known it was coming. But, the inevitable realization that his life was no longer his own was still painful. Three demons were gone and now he was being drawn into pursuing a fourth. Welcome to your mission, he thought.

PART 5

RESIDUAL UNIDENTIFIED FLYING OBJECTS

Although the vast percentage of UFO reports can be explained naturally, a residual percentage remains. These are called "RUFOs" for "residual UFOs". If only 1 percent of UFO reports remain unexplained, the number of RUFOs sighted over the last five decades could range in to the tens of thousands, if not many more.

Kenneth Samples

RUFOs are both real and nonphysical, and as such, they manifest specific characteristics. Examining these characteristics leaves the distinct impression that they have an intelligence and a strategic purpose behind them.

Hugh Ross

Chapter 27

Dr. Cassandra Sebastian stood in front of the closed door. A padlock hung from a roughly placed latch. She had showered in the boy's room and put on one of his pair of shorts and a tee shirt. Something called to her from beyond the door. She couldn't explain the calling. It was just there. Since childhood she had developed a sixth sense about these things. Objects of arcane antiquity called to her. She could sense them as she neared. This inexplicable ability was what had allowed her to be so successful as an archeologist. But, it had also led to her alienation from most of her friends as a child. And, her father. He told her it was because of her disease.

She reached out to touch the door. Dark red light exploded around her. She stood in the center of a bedroom and the walls were covered with blood. Not just blood, but letters and words in an ancient script scrawled in human blood! Screams echoed in her head and she collapsed onto the floor. Ghostly, ethereal shapes converged on her. She shut her eyes and put her hands over her ears. The screeching continued and hot blood dripped onto her face.

"Dr. Sebastian! Cassandra! Cassie!" Something jerked her up off the floor and she opened her eyes. Jonathan Steel stood in front of her. They were in the hallway outside the door. She gasped for breath and threw herself into his arms. She hung onto his body like a drowning person. Steel stiffened in her grasp but she didn't care. She had to push away the ugly, raw pain of the other world she had sensed. Tears ran down her cheeks and she was shaking. Finally, his arms encircled her shoulders.

"What's wrong?" He whispered.

"That room. What happened in there?" She managed to say as she looked up into his luminous turquoise eyes. His face was moist from his recent shower. His hair was wet. His breath smelled of wintergreen.

He abruptly pushed her away. "Someone died. Someone I was close to."

Sebastian regained her composure and wiped the tears from her face. "I'm sorry. Sometimes I have that kind of reaction to--" She shrugged. "Never mind."

Steel glanced at the door. "No one has been in there since the night of the attack." He looked back at her and his eyes filled with fire. "No one." He whirled and headed back down the stairs.

Renee sat at the kitchen table across from Arthur. Neither of them had cleaned up yet but they both sipped at coffee. Renee's cell phone dinged and she studied the screen. "That's odd. A text from someone to turn on my laptop and get online."

Steel hurried down the stairs. "Maybe it's Liz Washington. I've lost my cell phone and she was checking on Josh, Cephas, and Theo."

Renee had brought in her backpack and the semi-automatic rifle. She pulled her laptop out of the backpack and powered it up. It warbled indicating an incoming video chat. She accepted the invitation and a window opened on the screen. Staring back at her were two bloody, swollen faces.

"What is this?"

They were eclipsed by the smiling, hairless face of Anthony Cobalt. "Renee Miller, you see before you two of your distant acquaintances. Vivian tells me you know both Joshua Knight and Dr. Lawrence."

Steel roared in anger and spun the laptop toward him. "No!"

"Ah, Mr. Jonathan Steel. What a surprise to see you. You seem to have escaped my death pit in the old mission. Congratulations. You must realize I am never easily surprised. Reminds me of a story."

"Cobalt, if you hurt them, I'll kill you." Steel growled.

"I assure you Mr. Steel that as powerful and intimidating as Ketrick and Wulf were, they pale in comparison to me. I have been around since the Garden. I was Lucifer's right hand man. And, I've been planning this since then. Now, if you will kindly cooperate, this will all be over soon and your loved ones will be back in your hands. I have no interest in killing them. My plans are so much bigger than this world can contain." Cobalt raised his right hand and studied his fingernails. "Will you cooperate?"

"What do you want me to do?"

"There is an artifact. It is the last of three items I need to complete my grand plan. You have the resources to obtain it for me. You see, Dr. Miller is well aware of this artifact and I know that she can locate it for you. I want you to bring the item to me."

Steel glared at Cassie and she shook her head. Don't tell him we know about the Fallen Throne she said quietly to Steel and Renee. Renee paled and looked over her shoulder at Steel. He gently pushed her aside and sat before the laptop.

"What about the children?"

Cobalt lifted an eyebrow. "You mean the Children of Anak? They are very comfortable. Magan Celeste is seeing to their needs and her church congregation worships the ground they walk on. I will not harm them. You have forty-eight hours to bring me the Fallen Throne, Mr. Steel. Dr. Miller's phone has a number. Call me when you've found it." The screen went blank.

Steel slammed the laptop shut. "This can't be happening."

Knight stood up. "Jonathan."

"I will kill him! With my bare hands!"

"Jonathan!" Knight said again.

"What?" Steel shouted.

"Wasn't there a third person with Josh and Cephas?"

Steel froze. "Where's Theo?"

"Don't worry, honey, Theo will pull through." Liz Washington said over the cell phone.

Steel pounded his fist onto the wooden table on the deck outside the beach house. "What happened?"

"He was roughed up pretty good by Cobalt and Bile."

"Bile? That wimpy Bile?"

"Jonathan, Theo said Bile has a spiral tattoo around his right eye." Liz said.

Steel stood up and his heart raced. "What?"

"Yes, Jonathan. The thirteenth demon is back."

"Working with the tenth?"

"I doubt that. Thirteen never got along with any of the demons of the Council, remember? They kicked him off. He's probably working his own angle. And, I'm sure he saved Theo's life."

"Saved it?"

"Theo said Cobalt wanted to kill him and Bile asked if he could torture Theo first. He plastered him inside one of the walls of the cellar at the lake house."

"Can I talk to Theo?"

"No, he's in the emergency room under a sterile drape. Some pretty big lacerations they had to sew up. They had to give him a whopper of a dose of tranquilizers. Felt it was safer that way. When he wakes up, I'll get the whole story. I'll get him to call you. Now, what about Josh and Cephas?"

Steel told her briefly about the video call. Liz's voice was hoarse. "Oh, Jonathan! This is bad. Very bad. But, Cobalt is looking for the Fallen Throne?"

"Yes. Do you have any idea where this thing is?"

"You know I don't believe in coincidence, Jonathan." Liz said. "Yesterday morning, we broke through a wall off of the old church basement. There was a chamber built out of baked mud, primitive bricks put there by the descendants of the Inca who built the altar of the spiral eye. There were many drawings in the chamber. And, Jonathan, one of them was the Fallen Throne."

Steel motioned to Renee and Cassie to join him on the

patio. "Let me put you on speaker phone and repeat what you just told me."

Liz repeated it and went on with her story. "I've always thought the throne was just a rumor, a legend."

"Like the Lolladoff Disc." Cassie said. "Sorry, Dr. Washington, this is Cassandra Sebastian."

"If I could make the sound of a whip cracking, I would." Liz said. "On the wall of this chamber, there is an etching of the Infinity Disc. I can text you a photograph of the drawings and etchings if you like."

"Yes, that would be helpful." Renee said.

"But, here's the strange thing, Jonathan. The Fallen Throne is drawn sitting inside a chamber with another object."

"The Bloodstone?"

"No, the Ark of the covenant."

Steel looked up at Renee and Cassie. "You mean to find the Fallen Throne, all we have to do is find the Ark of the Covenant?"

"Jonathan, this is no coincidence. God is directing our path; your path to find the Fallen Throne. This is all meant to happen so you can save Cephas and Josh."

"Well, I wish the directions were a bit clearer. Do you have any idea where we can find the Ark of the Covenant besides in a United States Government warehouse?"

"I'm sorry, Jonathan. My forte is ancient languages. Maybe Dr. Sebastian can tell you where to find it." Liz said.

"Thanks. Liz, keep me posted. If you have any more information, call me back on this phone. It's Renee Miller's." He ended the call and turned to Sebastian. "You know something about where we can find the Ark of the Covenant?"

Renee appeared beside her. "If Cassie can beat me to the punch she would have done so already. Archeologists have been looking for the ark for over two thousand years."

"I know where the ark is." Sebastian said quietly.

"What?" Steel looked at her.

"Oh, come on, Cassie! If you knew where the ark was, you

would have had it on your show during the sweep weeks." Renee said.

"That's the problem. You can't get to it. It's guarded by priests."

"No!" Renee pointed a finger at her. "We've been through this before. It's not in Ethiopia."

"Renee, the Ark is there."

"How do you know?"

"You remember that guest I had on the show, the one who claimed to be a former priest?"

"I don't watch your show, Cassie. It's all a crock!"

"I thought this guy was faking it. But, the next day, he died of a heart attack in his hotel room." Cassie drew a deep breath. "I paid a pathologist to do an autopsy. His coronaries were wide open."

"So?"

"He left me a note, Renee. Passed it to me in the green room. The note said he was in danger and if he died under mysterious circumstances, I would know he was telling the truth. The note was in an envelope that contained a blurry photo of the ark. It's there, Renee."

"Where are the two of you talking about?" Steel said.

"Axum is Ethiopia's oldest city." Cassie said. "Axum dates back some 2,000 years to when it was the hub of the Axumite Empire. The Queen of Sheba made it her capital 1000 years before Christ. This capital city was the first place in Ethiopia to adopt Christianity. The Ethiopian Orthodox Church was founded there in the fourth century and Axum remains the holiest city of the Ethiopian Orthodox Church."

"In Axum, there is a church, the Church of Our Lady Mary of Zion. It is the most holy church of the Ethiopian Orthodox Church." Cassie's eyes filled with wonder. "I went there once. In the center of the church grounds sits the chapel of the Ark of the Covenant. Inside, is a red curtain that hides the Ark. A man stands in the doorway to the chapel. He is the guardian monk and it is an appointment for life. He never leaves the grounds. He must stay there until he dies. The reason the monk on my show died is because he abandoned his duties. You see, to gaze upon the Ark and not respect it is to sicken and die." Cassie reached out and placed a hand on Steel's arm. "If the throne is with the Ark, then we have to go to Axum and we don't have much time."

Renee's cell phone rang and Steel handed it to her. She answered. "Who? Yeah, he's here." She looked at Steel. "Do you know an FBI Special Agent Franklin Ross?"

"Of all the people to call me, I never expected you." Steel said.

"I hoped I'd never talk to you again, either. Especially after having to track you down through Dr. Washington." Ross' voice hissed over the speaker. "But, you owe me big time, Steel. Let me guess. You're in deep with another demon and you've lost the teenager and the old man. Again. Right?"

"I don't have time to play games, Ross. What do you want?"

"I know where they are. Interested?"

Steel listened as Ross told him about seeing Josh and Cephas. "So, they're at this Church of the Enochians with Magan Celeste and her congregation?"

"And Cobalt and your old girlfriend."

"Vivian!" Steel hissed. "What's your plan?"

"I'm trying to link Magan Celeste with laundered money. But, I can't ignore kidnapping, Steel." Ross said.

"Ross, if you storm that compound, Cobalt will kill Josh and Cephas. And, there are thirteen innocent children Magan Celeste has kidnapped."

"What do you suggest I do, Steel? Sit on my rear end and ignore this? I'm calling you out of a favor. And, I suspect there's some kind of paranormal angle here and I don't want that on my record again."

Steel's mind raced. He was running out of time and he needed an army at his disposal. An idea began to form and he glanced at Knight sitting inside the beach house at the dining table. He wasn't going to like it. "I have an idea. There is a military contact interested in these children. I could get them involved if you are willing to work with them. Are you interested?"

"Do I have a choice? If I don't agree with you, you'll track me down and break my nose again." Ross growled.

"Let me write down this number." Steel scribbled the cell phone number on a piece of paper. "I know who to contact, Ross. If you could hold on a few hours."

"Out of the question!" Knight stood up and paced across the kitchen.

"We don't have much choice. Josh and Cephas are at this UFO church compound somewhere in Nevada. And, we have less than forty-eight hours to find the Fallen Throne or Cobalt will kill them."

"So, storm the compound and stop Cobalt." Knight growled and the tiny red stone glowed beneath his tee shirt. "No! I didn't mean that! Cobalt would see us coming miles away."

"We divide and conquer. You and Renee will contact Robert Miller and work out a deal to confront Cobalt. Cassie and I will go to Ethiopia and find the throne. If it all works out, we may never have to give the throne to Cobalt."

"You called me Cassie." Sebastian said.

Steel blinked and tried to ignore her remark. "I know you don't want to work with Miller, but if he thinks he can get his hands on the children, he will bring everything he has to bear against the compound. We don't have much choice. Whatever Cobalt has planned for the children of the Bloodstone, it has to be much worse than what Miller has planned."

"He's right, Arthur." Renee reached out and took his arm. Knight slumped.

"I know. Either way we're making a deal with the devil. How do we contact Miller?"

Chapter 28

Vega fidgeted with her white jumpsuit zipper while the rest of the children milled around behind her. After the ships had departed from the platform, Magan Celeste had taken Vega and her friends to a dormitory attached to the stage. The dormitory was one building among many in a compound perched on the edge of a huge crater. Outside the crater, the desert stretched away from them into the far distance. Inside the crater a sphere the size of a small house hung over the center of the crater suspended by six huge struts arising from the edge of the crater. She looked through a huge window of the dormitory down into the crater. The far edge was over two miles away and a vast collection of electronic tiles covered the bottom of the crater.

"It's a radio telescope." Altair said as she joined Vega at the window.

"Magan Celeste said something about the Church of the Enochians. This doesn't look like a church to me."

"Beats me." Altair said. "Mother wouldn't lie to us."

The door leading from the dormitory back onto to the platform opened and Mama Celeste appeared, resplendent in an iridescent robe.

"Children, your time has come. The Enochians await you." Her face was flushed and she motioned behind her. "If you will join me on the stage, please."

The other children streamed through the door, but Vega stayed put. "Vega, please come with me. They are waiting." Magan Celeste motioned to her.

"To what? Worship us?" Vega said. "You said something on the ship about worship."

"Yes, they realize how special you are. They only wish to get to know you and to understand your power and your abilities." Mama Celeste's voice was soft and pleading. "Give them a chance, please?"

"And, then I can see my mother?" She looked Mama Celeste fully in the face.

"Of course." She said. Vega watched her pupils dilate. She listened careful to the rhythm of the woman's heart. It accelerated. Altair had been wrong. Her mother would lie to them!

The back of the platform had been closed off to the desert and Vega joined the other twelve children standing in front of chairs facing the forward curtain. Magan Celeste smiled at them and nodded to someone off stage and the curtain lifted.

Over two hundred people clad in white jumpsuits just like theirs sat in chairs. At the sight of Vega and the children, they rose as one to their feet.

"My Enochians, I present to you the Children of the Bloodstone." Magan Celeste shouted.

The people pushed aside their chairs and fell onto their faces. As one they began to chant in a language Vega did not understand. She glanced at Altair. Deneb shrugged in confusion.

"What are they saying?" He asked.

"Gibberish." Altair answered.

Magan Celeste began to sway in time to the chanting and Vega rolled her eyes as she slumped into her chair. "This is silly." She whispered.

Back and forth Magan Celeste swayed as she chanted in unison with the children. On either side of the stage, two short ivory columns with silver bowls allowed some sickly sweet smoke to taint the air. Vega's head hurt from the smoke. "Okay, I'm bored."

Magan Celeste shouted. "Be quiet for he is coming!"

The people across the empty floor of the auditorium fell

silent. Slowly, they sat up into a cross legged position and their eyes were riveted on Magan Celeste.

"Who is coming?" Deneb shouted. The other children laughed and Magan Celestial whirled in anger.

"Enough from you." She growled. And, then the expression on her face softened and she smiled. "Sorry for that outburst. I didn't mean to be angry with the Children of the Bloodstone. Or, as you will be known to your worshippers, the Children of Anak." She glanced over her shoulder. With that word, the assembled worshippers began to chant.

"Anak! Anak! Anak!" Vega listened to the words as they echoed throughout the room and a chill ran down her spine. These people actually worshipped them like they were some kind of gods. But, she knew there was only one God and Mrs. Donnelly had taught her and the other children all about God. These people were not worshipping the true God. Magan Celeste had fooled them all in one giant game of make believe.

Magan Celeste motioned for silence and she turned away from her followers. She walked slowly down the line of chairs. "My children, there is a wise one far greater than any person on this planet. This seeker of wisdom contacted me ten years ago on board an inter-dimensional ship from the galaxy of Andromeda."

Deneb giggled and Magan Celeste stopped abruptly. "You laugh, but in just a few moments, you will get to meet this spirit guide for yourselves. He is coming now across the vast, cold emptiness of space in an astral projection to speak to you and the assembled children about the future of his followers."

"What does he look like?" Deneb asked.

"You will not see him in his true form, for to look upon this creature is to suffer great pain and death. Instead, he will speak through me."

Vega's mouth fell open. Before Deneb could say another thing, Magan Celeste jerked and quivered as if in the grips of a great power. She stumbled back and forth across the stage until she froze. She threw back her head and straightened. It was as if she had become someone else. Vega leaned back in her chair in fear. Waves after wave of evil cascaded off of Magan Celeste. She turned and glared at Vega and her friends. Here eyes glowed with an unearthly green sheen.

"Ah, the Children of Anak." She rasped in a masculine voice. "I see that this assemblage has triumphed in bringing together those who will define the future of this galaxy."

Behind her, the Enochians rose to their feet. Murmuring and amazement filled the air. Mama Celeste turned back to them. "My faithful followers, I am Rudnalik, spirit guide from the planet Phlenum in the Andromeda galaxy. It is my wisdom that Magan Celeste follows for I bring you greetings in the name of our galactic realm. I bring you a promise of peace and harmony with all of creation. My faithful servant and channel, Magan Celeste has been my vessel for lo, these many years. And now, our plan to unite you with the Children of Anak is complete. Soon, you will journey beyond this world to a kingdom of unprecedented peace and prosperity." With that, Magan Celeste raised her hands into the air. The followers began to chant "Rudnalik! Rudnalik! Rudnalik!"

Vega felt a warm hand on her shoulder. She looked behind her into the eyes of Mrs. Donnelly. "Mrs. Donnelly! We wondered where you were."

Mrs. Donnelly leaned close to Vega. "Now is the time that I will have to leave, Vega. But, remember, I will be watching and I will not let any harm come to you. You must take charge of the children once I am gone. Understand?" Her bright eyes burned with passion.

Before Vega could reply, Mrs. Donnelly strode across the stage. The assemblage fell silent. "Enough of this nonsense, Magan Celeste. In the name of my God I command your demon to submit. I command Rudnalik to be silent."

Magan Celeste jerked and stumbled away from Mrs. Donnelly. Her body contorted in spasms. "Leave us, Donnatto. These have chosen."

"Not the children, Rudnalik. They have chosen God. You are commanded to leave." Mrs. Donnelly said.

Magan Celeste went into spasm again and gestured to the audience. "They follow me. They want my power. They want my presence. Behold!" With a gesture, Magan Celeste filled the air with light. Sparks of red and green cascaded out over the crowd and settled on them like pixie dust. Slowly, the assembled followers began to drift upward into the air. Their faces filled with wonder. They were flying!

"Enough!" The voice thundered across the room. A tall man appeared in the back doors of the auditorium. Like Magan Celeste, he had no hair. The assembled followers fell to the floor in a murmur of fear and confusion. He strode through them as if they were dust balls beneath Vega's bed, up the stairs, and onto the stage.

"I am Anthony Cobalt, children. It was my ship that brought you here and saved you from Major Miller." Vega nodded as she recognized the voice that had spoken to Magan Celeste in the ship's control room. Cobalt paused in front of Magan Celeste and his gaze slowly fell over the children. "I am most displeased with your choices, Rudnalik. Go back to the Eagle's nest and await my command. Depart. Now."

Magan Celeste spasmed again and then shook her head in confusion. She looked around and her gaze settled on Mrs. Donnelly.

"This is all your fault!" She rasped in her normal voice.

"I'm not the one deceiving these people." Mrs. Donnelly gestured to the audience.

"I am trying to build a following here for Mr. Cobalt. This church of the Enochians will follow him wherever he leads."

"Oh, be quiet Magan!" Anthony Cobalt shouted. "You've messed up things enough already. Why didn't you pay attention to that small voice inside of you? Huh?" Cobalt glared at her. "Kidnapping the children before I told you to was stupid and has drawn unnecessary attention from the military. Now, I will have to speed up my plans and when you take shortcuts, bad things happen."

Magan Celeste glanced at her followers. "Not in front of my followers, Cobalt. We had a deal."

"Then I suggest you stop the cheap theatrics. Get them ready for transport. We're leaving this place shortly and I will need them at the final Event." Cobalt said. He ignored Magan

Celeste and glared at Mrs. Donnelly. "I wondered where you have been! You haven't tried to stop me in years."

"I'm here protecting these children."

"No need to protect them anymore. They belong to me."

Mrs. Donnelly shook her head. "I don't think so. They are still innocent and still under the protection of the Father. You cannot touch them."

"True. I was hoping to wait until they were beyond the innocent years your master has so arbitrarily established. In a very few years, they will be able to choose for themselves who they will serve. But, Magan Celeste has forced my hand. And, the coming Event provides me with the best opportunity to carry out my plan. So, I will have to take them from you and watch over them until they can choose for themselves."

"You cannot do that." Mrs. Donnelly said.

Cobalt nodded and turned his back on her. He motioned to the back door. "Then, I shall have to show you who is in charge."

Vega watched a man shove his way through the double doors holding onto a teenage boy and an old man. The old man stumbled and his hair was wild and tousled. The teenage boy wore a look of utter defiance. Cobalt turned back to Mrs. Donnelly.

"You will leave now or I will kill one of these hostages. I don't care which. I will let Bile choose who to kill."

Mrs. Donnelly's hands fell to her side and she turned slowly. Her eyes met Vega's. "Very well. I will leave for now. But, I will return."

In a flash of light, Mrs. Donnelly disappeared from view, leaving the children stunned and speechless.

Chapter 29

Josh gasped when the woman on the stage disappeared from sight. "What's going on, Uncle Cephas?"

Bile squeezed tighter on his neck. "Shut up."

Anthony Cobalt glanced at them from the stage. "Take the boy to a dormitory and wait for me. Then, take Dr. Lawrence to the maintenance building."

Bile shoved them both through the open doors back out into the harsh sunlight. One of the worshippers from inside followed Bile out. "They told me to help you."

Bile glared at the young man in a white jump suit. "Then, take the old man over to the tech room. You can handle him, can't you?"

The man nodded nervously. "Yeah. Come on, grandpa." He took Cephas by the arm and led him across the compound. Bile prodded Josh across the compound toward an adjacent building.

"Where's your friend?" Josh asked.

"Who?"

"Thirteen? Is he messing with your mind, yet? Dude, he'll screw up your head."

Bile laughed and pushed Josh through a doorway into a building. "He's off with more important work than keeping an eye on you."

"Who is?" Anthony Cobalt followed them through the door. Bile swallowed.

"Jonathan Steel. I said he wasn't keeping an eye on Josh very well."

Josh blinked. Why had Bile lied? Cobalt motioned to the

door. "Get out of here and go take care of Lawrence. I can handle this."

Bile disappeared through the doorway and Cobalt studied Josh like he was an insect captured in a jar. "Now, the ground rules are very simple. If you so much as cause the smallest amount of trouble, I will crush the old man's head. You'll find clothing over there on the shelves. I suggest you shower and change. You stink."

"At least it's not fire and brimstone." Josh held out his hands. "Do I have to keep these ties on?"

Cobalt laughed. "Fire and brimstone? Really? My but you have such a quaint appreciation of my side of reality." He motioned with a finger and the ties disappeared. "I think you'll find if you try and leave this compound that is what will await you. We're surrounded by hot, desolate desert terrain. Once you're dressed, you can report to the mess hall for food." He stepped out into the heat.

Bunk beds covered the floor of the dormitory in a huge open room. No luxuries here. After all, it was Magan Celeste's "new world" that would afford luxuries. The lingering fragrance of the sickly sweet incense from the auditorium was the only thing out of place. As he inhaled the fragrance, Josh's head had begun to throb. There was something in the incense that affected the brain. Maybe it was some kind of chemical that made the masses more prone to hypnotic suggestion. Maybe he could wash the odor from his skin and hair.

The hot water felt wonderful and he used the communal shampoo and soap to scrub his body. Then, he stood under the shower head and let the hot water pound his neck and shoulders. Soon, the cloying odor of the incense was gone and his head had cleared. Afterwards, he turned the knob to cold. He yipped as the cold water bathed his head and shoulders. At first, it was shocking but it began to feel good as it washed away the heat. When he was finished, he toweled dry and pulled on the white jumpsuit. When he walked out into the open dormitory, it was filled with members of Magan Celeste's congregation. They were all standing next to their bunks, eyes riveted on him. Silent and eerie, they stared at him as he moved around the periphery of the room. Most of the people were older than him, probably below the age of thirty. Josh shook his head. These poor souls were deluded by Magan Celeste's alien 'gods' and her promise of a new world. Hadn't it always been this way throughout history? Josh was glad he had discovered the real and lasting Truth and was no longer misled by the whispers of evil. He scooped up a handful of granola bars and a couple of bottles of water from the snack table and paused at the door. Eyes filled with anger, they stared at him in rapt silence. Josh nodded and stepped through the door back out into the heat. No one tried to stop him. Now, to find Cephas.

"Okay, all of the other buildings look the same on the outside." He said to himself as he chewed the granola bar. "Now, there were about forty beds in the dorm and Magan Celeste seems to have over two hundred followers so that means she needs at least five dormitories." In all, he counted eleven buildings including the "church" of the Enochians. One building had to be an eating hall. "Dude, there has to be some kind of administrative building with classrooms. So, start looking."

On the third try, he found the administrative building. It had offices and there were happy, deluded Enochians manning phones and computer consoles. They were selling hygiene products! Josh peeked into several offices and saw the products laid out on a table. One of Magan Celeste's followers was on a phone in each office pitching their sales of these products from "out of this world" promising "celestial cleansing" and "heavenly stress relief". Josh ignored them and made his way around the periphery of the office building and he ran across a different room in the back. It was larger than the offices and occupied a third of one end of the building. Could this be the "tech" room? He knocked on the door and waited. The girl that opened the door towered over him by almost two feet. The door frame reached her chin and she had to lean over to see him.

"Who are you?"

Josh's mouth fell open and he gasped. "You must be Vega!"

"You know me?"

"I know your mother."

Vega's eyes widened and she glanced out at the empty hallway. She grabbed Josh's arm with an iron grip and jerked him into the room. It was brightly lit and filled with toys. Only these toys were three times their normal size. Twelve other children the size of Vega stood up around the room and looked at him. The world tilted on its axis and for a fleeting second he was the child surrounded by giants. They all appeared perfectly normal and proportioned. They looked like ordinary kids only twice their normal size. No, three times, he corrected himself. When he had seen them on the stage, they had appeared almost normal.

Vega dragged him to the center of the room. "Hey, everybody, he knows where my mother is."

They bombarded him with questions all at once asking him about their mothers and where they were and when they could see them. Josh put his hands over his ears and had to shout.

"Wait a minute! Dudes, do you think you could calm down? I mean take a chill pill or something?"

Vega put her finger to her lips and they all quieted down. "Remember what Mrs. Donnelly told us. She said someone would come to help. Are you that someone?"

"Maybe. Unfortunately, I'm a prisoner just like you."

"We're not prisoners." One of the boys said. "We're guests. Those Pinnochian people bow down to us. It's weird."

They all started talking at once again and Josh pulled Vega aside. "Listen, is there somewhere you and I can talk alone? In private?"

"You mean like a time out?" Vega said.

Josh looked into her innocent face. He had to remind himself she was only eight years old in spite of her size. "Yeah, a time out would be cool."

Vega pulled him across the floor to the far end and stepped into another room. She closed the door behind her. It was a smaller room with a window that looked out onto the compound. Four desks with computers sat in the room. Josh ran to the nearest computer and started tapping at the keys.

"It's no good. You can't get on the Internet." Vega said. "We're not connected. There are just games and school lessons."

Josh frowned and sat down in one of the oversized chairs. He was the incredible shrinking dude. "Great! I've got to get ahold of Jonathan."

"Who's he?" Vega settled into a chair at the desk next to him.

"My guardian."

Vega smiled. "I have a Guardian, too. But, I don't know his name. Do you have a father?"

Josh paused before answering the question. He hadn't thought of his father in a long time. And, now, according to Uncle Cephas, his father was alive. Was it true? If so, why hadn't his father spoken to him in the years since he supposedly died? He frowned and glared at Vega. "My father is dead. Yeah, dead. Jonathan is taking care of me now. And, my Uncle Cephas."

"I think I met my father in the woods. He was trying to rescue me. He has never told me he is my father, just my Guardian." Vega frowned. "But, my invisible friend told me he was my father."

Josh sat forward and looked at her. "You mean Major Miller?"

Vega's face twisted in disgust. "No! Major Miller is mean! He's trying to take us away from our mothers! He works for the military!"

Josh blinked in confusion. Oh yeah, he thought, Renee had divorced Miller and probably never told Vega he was her father. But, then who did she think was her father? "I don't understand. Who is your father?"

Vega shrugged. "I never knew him but he has always watched out for me. I know. In my heart, I know. He's like God. God is my Father, too and He watches over us and sends us a guardian angel and protects us from evil. God loves us and takes care of us even though we can't see Him all the time. But, He's there. Just like my real father. He's there and I can sense Him even though I've never seen Him."

"Yeah, fathers are like that, aren't they?" Josh looked away. Had his father watched him from a distance?

"Was he nice like Mrs. Donnelly or mean like Mr. Cobalt?" Vega asked.

Josh looked up into her big eyes and soft face. He thought back to the years he had spent with his father. They had been good. When he was around. "Not all the time. Sometimes, he was a little mean because he wasn't perfect, you know. But, he loved me." Did he? Josh sat forward in confusion again. If his father loved him, then why had he stayed away all these years?

Vega picked at her nose and sucked on her teeth for a second. "You know, I think that sometimes mean people have mean mothers and fathers. Think about it. If your father is really mean and evil like Mr. Cobalt then that would mean you would think Father God is that way, too. Is that why some people don't believe in God?"

"I've never thought of that. For a while I hated God. I guess it was because of my father."

"So he was more than a little mean?"

"Well, he was preoccupied with something, Vega. Before he disappeared he and my mother had lots of fights. I don't know if they loved each other anymore."

"Well, Mr. Cobalt is not just mean, he's evil. And, if there's evil around then there must be good around. So, that proves that God is real. Why can't people see that?" Vega cocked her head at an angle. "I wish people could see God like I do." Vega leaned toward him. "I used to see invisible monsters and invisible friends all the time until I figured out that Mrs. Donnelly is really my guardian angel. And, if I have a guardian angel, then God sent it. Right?"

"I think you're right, Vega. I had a guardian angel once. He chased away some really bad people." Josh stood up and crossed to the window as he recalled the angel clad in green scrubs on the river bluff in Lakeside. "I wish he were here right now."

"You can borrow my angel, Josh. I just hope our guardian angel takes care of my father." Vega said.

The door opened to the room and Anthony Cobalt stepped in. His evil rolled across the room like a wave and engulfed him. Josh gasped as his thoughts clouded.

"Vega, there you are. Magan Celeste needs you and the children. Please go to her." He pointed to the door. Vega looked down at Josh and Josh nodded.

"Go ahead. I'll be fine."

She leaned over. "Remember our angel."

Vega left the room and Cobalt closed the door behind her. "It didn't take you long to leave the dorm behind."

"You told me I could go anywhere." Josh glared at him with as much defiance in his voice as he could muster. But, the defiance didn't cover up his fear. A strange paralysis filled his limbs and his tongue thickened like what he experienced with the incense. He shook his head and pushed away Cobalt's evil influence.

"There was a remote village far removed from civilization." Cobalt raised a hairless eyebrow. He sat in the huge chair. "Its occupants had never seen a civilized 'white' man. They lived in peace and harmony. And then, one day a strange roaring noise came from the heavens. The villagers ran out and looked up into the blue sky and beheld a flying god traversing the clear sky. The god had wings and roared with a mighty sound. The villagers fell on their faces in wonder and awe." He stood up and crossed to the window. "And, fear. The god disappeared. The next day, they waited and he did not return. Nor, on the next. Finally, the shaman of the village decreed that an image of the god must be erected in the center of the village. No cost was spared. Huts were demolished. Hours were spent constructing an idol to their new god. Women and children starved because the men refused to hunt until their idol was finished. Weeks passed and the idol was completed and life returned to normal. But, their god never returned. And, so the shaman assumed control of this new religion and rose to power and fame. He soon became the leader of the village along with his children."

"Sixty years later, a group of explorers discovered the remote village. Half of them died in an ambush. The other half were taken to the village square where they were sacrificed to their god. One of them escaped and led a well armed group against the village. They never knew what hit them. Everyone man, woman, and child in the village was killed in retribution for the murders of the man's colleagues. And, much to their amazement, they discovered in the center of the village an almost exact replica of a biplane constructed from rare jewels, fine metal, and stone."

Cobalt turned away from the window and smiled at Josh. "Wondering why I told you this story?"

"Time for my nap?" Josh said.

Cobalt frowned. "Don't you see, Josh? I am a god to these people. They will kill for me. And, they will die for me. They have no idea of my true nature." He walked toward Josh. "And, they have no desire to know the truth. They are content to reside in their delusion. Because, Josh, sometimes a familiar delusion is preferred to a more difficult truth. And, even more important than that, the Children of Anak are like gods to them. And, they will gladly die for the children, too."

"What do you want with Vega and her friends?" Josh said.

Cobalt's lavender eyes glowed with power. "They will become my children. In my new world, they will be my conquerors, my army. I managed to gather their kind behind my power in the past. You've heard of the Philistines, haven't you?"

"Yeah, in the Bible."

"They were giants in the land." Cobalt said. "And, we were going to conquer the world."

"Until a little boy with five stones stopped you, bro." Josh said. And, with that thought, the confusion melted away.

Cobalt jerked as if he had been stung. He glared at Josh. "David and Goliath! Impertinent little runt! I should have killed him when he was born."

"You couldn't have killed him. He was protected by God." Josh's strength was beginning to return. "Didn't see that one coming, did you, bro? Thought you had it won because of your giants, didn't you? Well, God is so much bigger than you and your lame demon, Cobalt. He has someone else he is sending who will be your David. His name is Jonathan Steel."

Cobalt lashed out with his hand and grabbed Josh by the neck. Suddenly, his features twisted in pain and he released Josh. He backed away and looked at his hand is if it were burned.

"Cat got your tongue?" Josh said as he rubbed his neck.

"I see I was mistaken in allowing you to regain the strength of your faith." Cobalt glared at him. "I could have snatched your tongue from your mouth and swallowed it before you knew it was gone."

"I don't think so." Josh stood straighter. "You can't even choke me. You see, I'm not as confused now as I was in the basement of Ketrick's house. Probably had a little concussion from Bile. Well, I got my mojo back."

Cobalt rubbed his hand and regained his composure as he slid out of the chair. "Did Vega tell you about her father?"

"What?"

"He tried to rescue Vega. And so did Jonathan Steel." Cobalt smiled. "In fact, they were together, Josh. They were working together. He was the third person in the cave with Steel and Sebastian. I thought I had fried all three of them, but it seems they managed to survive." Cobalt stepped closer. "Your exalted Jonathan Steel hasn't been honest with you. He's been working with this man and he has told no one."

"What man? You're not going to tell me my father is alive again, are you?"

"Yes, Josh. Your father never died. He hid from the world. He hid from you. And, he showed up to rescue Vega and he failed. And then, he tried to help Jonathan Steel rescue Vega and he failed." Cobalt hissed and then glanced out the window. "I wonder if he'll try and rescue you now? Probably not. He doesn't care about you, Josh."

Josh shook his head as confusion began to cloud his thoughts again. "But, Jonathan would have told me."

"Oh, you think so? Why? If he told you, he'd have to give you up to your real father. Think about this, Josh. Your real father ran away and hid from you and your mother when you needed him most. Now, he's back and he hasn't even bothered to let you know. And, Steel discovers your real father is alive but doesn't tell you." Cobalt clasped his hands behind his back. His luminous eyes glowed with evil. "Why is he hiding that fact from you, Josh? Have you thought about that? Jonathan wants to keep you for himself. He doesn't want you to go back to your father. Of course, why would you want to go back to a man who doesn't love you anymore? And why would you want to stay with a father figure who lies to you? You're better off without either one of them, Josh. Face it, you are alone in this world. Again!"

Josh looked into Cobalt's intense gaze. "No, bro, they wouldn't do this to me!"

"They did, Josh." Cobalt stepped closer. "It shouldn't have surprised me this would happen. Your father was always impulsive especially when we worked together."

Josh wiped his eyes. "What?"

"Your father worked on a special project for Major Miller and I helped him."

"That can't be."

"Oh but it can be, my boy. You see your father needed help of the, uh, supernatural kind to complete his task. He had to do something to please his superior, Major Miller. Or course, he wasn't a major at the time. Let's see, that was eight years ago." Cobalt placed a finger on his chin as if deep in thought. "Oh yes, you and your mother were gone to California. You were gone for two weeks. Yes, your father was very busy those two weeks." Cobalt straightened and started for the door. He paused just before leaving. "Oh, by the way. Another thing you've never been told. Vega is your half sister."

"What?"

"You have the same father, Josh. Consider that. Now, why would he spend the last few years watching over Vega, and not you? I wonder who he loves the most? Now, what was it you were saying about a rescue? My dear boy, there's nobody left who cares about you enough to rescue you. You're all alone. Forever!" Cobalt closed the door on the room. Josh felt like he had been kicked in the gut. His father was alive. Jonathan knew and had never told him. And, Vega's mother was Renee? And Vega's father was the same person as his father? That meant, his father and Renee? Oh, God, what is going on? He screamed it aloud and then all he could manage were sobs.

Chapter 30

Darkness had fallen by the time Josh recovered enough to stare out the window at the clear desert sky. Stars filled the emptiness with a grandeur he had never imagined while growing up in Dallas. Even through the glass of the window, they were amazing. A meteor soared across the starry landscape and blazed in fleeting glory as it disappeared on the horizon where a new moon was beginning to glow. He was spread out on the floor of the computer room in total misery. So many lies. So much confusion. Who could he believe anymore? His father was alive? And, his father had worked with Anthony Cobalt? Worse, his father had an affair with Renee Miller? And they had a daughter? And that daughter was Vega? How could he have been so blind? Of course, he had only been nine that year his mother took him to California. Later, he realized she had gone to see a medical specialist. And, he had been eleven when his father had died; wait, disappeared.

Josh recalled those years leading up to his eleventh birthday. His father and mother had fought constantly and he had known something was wrong. That was why he had rebelled. Now, he knew why they had been in such a state. Did his mother know about Vega? Did she know about Renee? He missed her so much! If she were just here for a moment. Why had he pushed her away? Why had he been such a jerk? He wasn't even himself when his mother had sacrificed herself to save him. His last conscious memory was of her hurtling into the fire to keep Robert Ketrick from killing Emile Parker in Lakeside. The pain was immense and it consumed him. It drew him down into a well of pain and sorrow. He closed his eyes and gave in to the fatigue and the pain.

"Josh, what are you doing?"

Josh opened his eyes. He wasn't in the computer room anymore. He was standing on the mountainside overlooking Albuquerque. He looked around and his mother stood there in a diaphanous gown. Her clothes seemed to glow from within and they stirred in a light breeze that carried the scent of gardenias. "Mother?"

She was beautiful and radiant. "Jonathan needs your help, son. You must wake up."

Josh stepped towards her. "But, what about father? All this stuff I've learned?"

Claire Knight smiled. "Oh, Josh, don't you know we worked all that out? I forgave your father years ago. We all make mistakes. Don't let that stop you from helping Jonathan. Don't buy into the enemy's lies. Remember, they are just lies. All lies."

Josh reached out for her. "You're just a dream, aren't you?"

"Yes. We cannot return to these dimensions, Josh. You know that. We talked about that when you were younger. Remember, you have always had dreams. God talks to some people in their dreams, Josh. You are special. That is your gift. Satan can influence your dreams, but he cannot talk to you in those dreams. These special dreams are from God. Your sons and daughters will prophesy, your young men will see visions, your old men will dream dreams. That's what the scriptures say, Josh. Well, you are not an old man, but you dream dreams that are sent from God. This dream you are having is from God. This is good. And, you can sense that, can't you."

Josh nodded and his heart ached. "Yes. I know evil when it is near. I know evil when it is in my dreams. So, what can I do?"

"Jonathan's life will be in your hands. You must help him."

"But, I miss you so much."

"I understand that, son. But, right now you must wake up. Now!"

Josh blinked and he was on his back staring out the window again. He sat up and looked around. The door opened and Bile and Vivian walked in escorting Cephas Lawrence.

Cephas wore his robot pajama pants and an undershirt and he smelled to high heaven. Bile jerked him over to the oversize chair.

"Uncle Cephas?" Josh stood up.

"Hang on, honey child." Vivian held up a hand. She looked around the room and shut the door behind her. "We only have a minute. Cobalt told us to bring Cephas here with you. I'm leaving on one of the ships to find the Fallen Throne. Now, what you need to understand is that I'm going to do my best to help Jonathan."

"What?" Josh backed away from her outstretched hand. "I don't believe you."

"You saw what happened in the basement of Ketrick's house." She said. "I told the golem to kill Cobalt. But, he is too powerful. So, you have to understand something. Cobalt is taking you to the Eagle's Nest with the rest of his sheep. He has something planned for the children of the Bloodstone. I don't know what. Cephas will stay here. So, you will be on your own. I will do my best to let Jonathan know where the two of you are."

"Enough of this." Bile growled. Josh glanced at his face. The tattoo moved around his right eye.

"The Council was very specific. Steel can't die. And I have to stop Cobalt. Isn't that what you want?" Vivian said to Bile.

"I don't care about Steel or Josh or the old man."

"You want me to do this, right? Then, let me do it my way. Why don't you take a walk if this is bothering you?"

Bile released Cephas and left the room. Vivian relaxed. "I know you don't believe me and frankly, I could care less. I know what I have to do and I'm giving you a fair warning."

"I believe you are growing soft hearted." Cephas said.

Vivian planted her hands on her hips. "No need to get

insulting. I'm leaving now so don't try anything stupid." She glanced at Josh one last time and left the room.

"What now?" Josh asked.

"We wait for an opportunity and pray." Cephas said.

Josh slumped to the floor. "I'm so tired, Uncle Cephas. I've heard all of these rumors about my father. How long have you known he was alive?"

"Jonathan told me the night he went to the school. Your father was there looking for the children. It was the first time either one of us knew he was alive." Cephas massaged the knot on the side of his head. "Before then, I had no idea, Josh."

"Why didn't Jonathan tell me then?"

"Tell you what? Hey, Josh, in the midst of being kidnapped by UFOs and little winged aliens guess what? I found your father and he is alive and well and he hasn't bothered to talk to you in six years?" Cephas rubbed at his moustache and he looked fragile and vulnerable in his undershirt and pajama bottoms. "Have you thought about what Jonathan is going through? Up pops your father who is supposed to be dead. And, Jonathan has made a promise to take care of you, not to mention fighting off a herd of vampires to save your life and a bunch of white eyed ghouls. Now, in one second all that is null and void. You go back to your father."

"I will not go back to the jerk."

"Don't call him that."

"Why not, Uncle C? Let's see. He had an affair with Renee Miller. He got her pregnant with a half sister I never knew about. And, he worked with Cobalt to make Vega the monster she thinks she is."

Cephas drew a deep breath. "Your father and Renee? Cobalt? They have been manipulating this behind the scenes for years? I didn't see this coming."

"Not only that, he abandoned my mother when she needed him most. And, he flew off into the sunset and left me without a father for six years. Six years, Uncle Cephas! Do you know how many times I wished my father would walk through my front door? Do you know how many times I looked at his car and his motorcycle sitting in my back yard and wished he would take me for a ride? Do you? Do you have any idea what it is like?"

Cephas was quiet for a second and then sighed deeply. He looked at Josh. "My father was a cipher during the Cold War. Do you know what a cipher was?"

Josh shook his head as his sudden anger began to abate. "No."

"He worked with the Army on codes and secret messages shortly after World War II. I was a young boy and all I wanted to do was to play in the street and throw my baseball. But, my father was very busy. That branch of the government was quite stressful. Father would come home late in the evening and stare at the wall. His mind was always working on the latest encryption technique developed by the Soviets. He would go outside when the sun was setting. In the alley behind our apartment building in New York City we would play pitch. Before he had thrown the second ball he would have another breakthrough. He'd drop his glove right there in the street and run up the fire escape to our second floor apartment to make notes. That would be it. I wouldn't see him again until I peeked my head into the closet he had turned into a small office to tell him good night."

"What happened?"

"It was my seventh birthday. Father had promised me he would take me to a movie. I wanted to see the latest serials at the theater. There was a bus that would come by and pick him up every morning. All of the ciphers rode the bus. It was a security issue. My Father waved to me from the street and I leaned out the window excited we were going to the movie that night after I came home from school and got all of my homework done. My mother was right there with me smiling and rubbing my back as he got on the bus. It almost made it to the corner before it blew up. Someone had sabotaged the bus to take out all the ciphers. We never found his entire body. Just bits and pieces."

Josh leaned back against the wall and closed his eyes. "I'm sorry. I didn't know."

Cephas looked at him and his tired, aging eyes were

filled with the weight of the world. "I know how you feel, Josh. Only, my father could never come back to me. Your father can. Whatever his reason for doing what he did you need to forgive him."

Josh sighed. "He wasn't always the best father. He never played ball with me. He was always working on his computer at night."

Cephas patted Josh's leg. "There is an interesting book that looks at the lives of famous men who claimed to be atheists or skeptics. In every case, the fathers of these men were universally abusive, mean, distant, self destructive, or absent. Sometimes, when I hear other men and women talk about not believing in the God of the universe who created us all, I begin to understand them. They are looking at God through the framework of their experience with their fathers. After all, we call God Father, do we not?"

Josh nodded. "What's your point?"

"God, the Father. He is the creator of the universe. From Him all things originate. He gives our universe structure and design and meaning. He gives our lives purpose and meaning. If your father had been different, you might hate God. You might equate God with all the anger and resentment you feel right now toward your father. But, you don't hate God, do you?"

Josh studied the old man's moist eyes and his sagging moustache. "No, I don't hate God. Sometimes, I don't understand why things happen the way they do. But, most of the time, it's my fault."

"Do you realize how unique you are, Josh? Most kids your age will never admit that they are to blame for their own problems. Who taught you that?"

"My father." Josh whispered.

"Exactly. Right now you are angry. And, there are many times I am angry at God. It is not because of my distant father. I knew he loved me in spite of his distance. I now understand that he did what he did out of a greater sense of right and wrong. He did what he did to ultimately protect me and give me a good future. I just couldn't see it at the time. Perhaps what your father did was to protect you. Have you considered that?"

Josh shook his head. "By pretending to be dead?"

Cephas leaned closer and nodded. "Exactly! Didn't Renee Miller say that she had someone she did not know who was taking care of the children in secret? Didn't Jonathan say that Major Miller came to the hospital looking for someone? Josh, I believe your father, for reasons we cannot understand, was in great danger. He knew that someone would try and hurt his family to get to him. The only way to protect you was for him to die. And that meant he would have to cut himself off from you permanently."

Josh blinked. "Uncle C, what you're saying is he loved me enough to die to protect me?"

"Precisely. Sort of like what God did when he became a man named Jesus and died to deliver us all from the tyranny of our selfishness. God's sacrifice was his son. Jesus' sacrifice was his life. Your father's sacrifice was to cut himself off from the only two people he loved. Think on that before you condemn him anymore. Wait and learn the whole story." Cephas patted Josh on the shoulder.

"If we ever get out of this room."

Cephas reached over and pulled Josh into a tight hug. "I will not let that fiend kill you, Josh. I will die before I will let that happen."

Josh fought back the tears and hugged the old man in return.

Chapter 31

It was all a blur. The helicopter to Pensacola air base. One of Renee's transport planes to Ethiopia. Now, the truck that bounced along rutted roads was pulling into Axum, Ethiopia. The stark, but beautiful countryside did not impress Steel. He was tired and discouraged. His few hours of restless sleep were filled with nightmares of Josh and Cephas at the hands of Anthony Cobalt.

The truck carrying them kicked up a cloud of dust. The road wove its way along the base of the Adwa Mountains. The city was spread out over a valley floor with scattered evergreen trees and low lying buildings. They pulled to a stop in front of an open expanse of space in the heart of the city. Surrounding them, tall obelisks reached toward the deep blue sky.

"We have less than three hours." He said.

Cassandra Sebastian thanked the driver in a foreign language and then motioned for them to get out of the truck. "We have plenty of time to find the throne."

Steel climbed out of the vehicle and studied the tall obelisks and flat spires surrounding them.

"So, this is the North Stelae Park." Sebastian pushed her mirrored sunglasses up into her hair. "Monoliths erected in memory of great leaders. That one in the middle was returned from Italy. Mussolini took it to Rome during World War II."

Steel noticed numerous stone monoliths lying broken and half buried by the grass and soil. "The throne is here somewhere?" Another truck pulled up the roadway.

"My other truck has arrived." Sebastian said.

"For what?"

"If I can't have the throne, then I'll have the Ark of the Covenant."

The old fury took him and he almost grabbed her. He wanted to shake the life out of her. "What? We're here to find the throne. I don't care about the Ark."

"You don't think I would come all the way to Ethiopia after learning that the Ark is here and not take it?" Sebastian put out her hand. "I don't have much hope left, Jonathan. See that tremor? It's getting worse. The Ark may be my only hope."

"Or, it might kill you. Isn't that what happened in the Bible to anyone who touched the Ark?"

Sebastian put her sunglasses back over her intense eyes. "I'll take my chances." She whirled to hop onto the new truck that had pulled up beside her in the middle of the road, but her way was blocked by a tall man in a multicolored robe. He had appeared out of nowhere. His skin was a black as the night and a tattered, but colorful turban covered his hair. He leaned against a dark wooden staff. The top of the staff had been carved into the image of an angel with four wings pointing toward heaven. A cherubim?

"You will come with me, Jonathan Steel." He said quietly.

"No, no, no, no, no!" Sebastian shook her head as she stepped back from the man. "I deserve to see. I do. You know it."

The man glanced at her and frowned. "We have spoken before and my answer is the same." The man gently pushed Sebastian aside and gestured to Steel. "It is time for you to face your destiny." His eyes were a deep brown and each pupil was rimmed in a circle of blue. "I am here for you, Jeremiah."

"My name is Jonathan."

"You have gone by many names. Jeremiah. Jonathan. Son. Lover. Killer. King. Hot Steel."

At the mention of that name, Steel dizziness overcame him and he remembered.

The restaurant sat on the apex of a small mountaintop

between the two ski slopes. A group of black SUVs were parked outside the restaurant on an access road. Lights gleamed from within. Hot Steel looked beyond the restaurant at the pale sky growing lighter as morning approached.

"Are you ready?" He asked Vega.

"I trust you." She said in her tiny voice.

They walked up a set of stairs toward the double door entrance to the restaurant. Wild flowers winked at him between the metal grate normally filled with the crushed snow of skiers. A heavyset man in a black suit nodded at him as he took Vega through the door to her fate. The interior of the restaurant had been cleared of its tables and chairs. The boy he had shot lay in the center of the room, his chest rising and falling. A similar sized girl with dark ringlets slept beside him. Major Miller was pacing back and forth. He halted and his eyes burned into Hot Steel's face.

"What is this, Hot Steel?"

"The last target, sir." Hot Steel said.

For a moment, Miller's eyes betrayed his recognition of Vega and then they took on a hard edge. "Why isn't she out?"

"I was alone. It would have been impossible to carry her up the side of the mountain. Besides, she's powerless." He said. Vega's hand tightened its grip.

Miller seemed confused and hesitant as he looked over his shoulder. A windowed conference room across the restaurant was hidden by shades. But, Hot Steel could easily recognize the shadow of a man in a wide brimmed hat smoking a pipe. Miller glanced at Vega.

"Put her with the others. And, stun her."

Hot Steel's face burned with the knowledge that his father was just thirty feet away. Abandon the girl. Go after his main objective. That is why he had done all of this. To find his father. To find his answers. And, perhaps to punish the man who had ruined his life.

"Mr. Hot Steel, are you going to help me?" She tugged his hand. "You promised."

Something broke behind his vision and the tension and hatred melted away. At his core, he was not going to be controlled by the resentment he felt for his father. It could not, would not determine who he was. If it did, then his father would have succeeded. He nodded. "Trust me, Vega."

"You shouldn't trust anyone, sweet pea."

Hot Steel looked up. The Captain stood on the other side of the room. He was shorter than Hot Steel and his face was shadowed by the wide brimmed Panama hat he wore. But, his bright turquoise eyes gleamed with malice. He puffed on his Meerschaum pipe and the fire painted his face in orange. "Your call sign is Hot Steel?"

"Yes." He managed hoarsely through clenched teeth.

"Well, it doesn't matter what you think. These three children were an experiment that went wrong. Some of them had powers and one of them doesn't. And, it seems those powers are not native to their altered minds as I was hoping. They're just channels for the evil around us." He walked up and poked the sleeping boy with a boot. "Between them, they killed fourteen well armed mercenaries, though. Now, if we could control those external forces, that would be different. But, these things," he nudged the boy again, "are just conduits."

The Captain turned to Miller. "We're done. Kill the children and take them back to the dissection lab. I've got a plane to catch." He paused and glanced at Hot Steel. "And, leave no witnesses."

Miller was clearly torn but he lifted the pistol from his holster and pointed it at Hot Steel. The sound of a fired rifle snapped across the room and Vega's eyes widened in shock. In quick succession, three darts appeared on her chest. She looked down and her eyes rolled back in her head. Hot Steel caught her as she fell. He glanced up. Stoneheart stepped out from behind a column.

"Hot Steel's not the only survivor. I'll do the job he can't finish. That's the way it always is with his type."

Miller glanced at Stoneheart and back at Vega. Stoneheart winked at Hot Steel as he reloaded his weapon with a new rack of real bullets. "Now, Major Miller, if you'll excuse me, I have someone that needs more than just stunning."

In rapid succession, Stoneheart began to shoot the four

bodyguards stationed around the room. Hot Steel dove forward over the children's bodies and pulled Vega beneath him. The four bodyguards fell in rapid succession and Miller slid beneath an overturned table for cover. He fired his pistol at Stoneheart, but the man dodged the bullets and ran toward the corner office.

Hot Steel could hear one of the SUVs tearing down the access road. The Captain had gotten away. Stoneheart dove through a window and shattered glass followed him out into the morning air. He jumped into the other SUV and started down the road after the Captain.

Miller slowly stood up and surveyed the carnage. He walked over to Hot Steel and held the pistol toward him. His hand was shaking.

"She may not be my daughter, but I was there when she was born. I was there when she was small. I can't kill her. Now, listen carefully. A man will show up shortly with an ambulance. It is to take the children back to the lab. You must stop that man and take the children yourself." He took a cell phone out of a breast pocket. "Call the number listed as Renee. Tell her where you are and she'll come get the children."

"Why?" Hot Steel whispered.

"Because I love her. She's my daughter. I can't help it. Now, I have to go and save my boss." He turned and ran from the room.

Hot Steel stood up slowly and looked down at the sleeping children. Would he ever love someone enough to sacrifice his life for them? Would he ever have children? If he did, would he be the kind of father the Captain had become? His hand closed involuntarily in anger on the cell phone and it shattered in his grip.

"Just put your hands up and don't move." He heard a voice behind him. He slowly raised his hands and turned around. He had expected to see Stoneheart. But, the man who stood before him was not one of the soldiers. He was dark haired and wore a white lab coat. But, he was more than just a scientist. His steady aim with the pistol spoke volumes of his true abilities.

"Why didn't you just shoot me?" Hot Steel asked.

"Information. Where is Miller?"

"He went after an assassin who came to kill the Captain. Why are you training a gun on me?" Hot Steel asked. Something wasn't quite right.

"I don't want you to kill the children." The man stated flatly.

"Then we're on the same side. I was about to call Renee."

"Is that why you broke the cell phone?"

"No, I was angry that it wasn't me going after the Captain." Hot Steel said.

The man seemed to waver. "I don't understand."

"Neither do I. Are you taking these children back to be dissected?"

"No. That attendant is back down the mountain somewhere in the bushes. I took his ambulance. And, I'm taking them to safety. It's my job to protect them."

"Then, you haven't been doing a very good job, have you?" Hot Steel whispered.

"No, I messed up and Miller got his hands on them. I got sloppy." He blinked and lowered the gun.

"I'm leaving now and you can take the children. Just keep them safe from monsters like Miller and the Captain." Hot Steel tossed the cell phone aside.

"Wait!" Vega stirred at his feet. She sat up and rubbed her eyes. Her voice was slurred as she reached for him. "You kept your promise."

Hot Steel reached out and took her hands and pulled her upright. She wobbled slightly and then hugged him. He felt awkward and then put his hands around her. "No more invisible monsters, okay? And, keep saying your prayers. God will protect you."

Her eyes were filled with tears. "And, you keep on keeping your promises. Promise?"

Hot Steel smiled in spite of himself. "Yeah, I promise."

Steel's vision cleared and he looked up into the face of

the man in the turban. "My son, that memory was important. Now, I must take you and leave this outcast behind."

"No you won't!" Sebastian screamed. She reached through the open window of the new truck and gestured to the driver. He handed her a pistol and she pointed it at the back of the man's head. "Now, you will take me to the ark --"

The holy man never turned, his gaze frozen on Steel. He gestured with an arthritic, deformed hand and the gun turned from a dark shiny metal to the color of brittle red clay. It crumbled in Sebastian's hands and fell to the ground. She swore and held up her empty hands. "How did you--"

"Her time has not yet come. She will see the Truth incarnate one day and stand on the place of the skull beside you, my son. But, until that day, she will reject the love of the Son. She is not yet worthy. But, you are. You must come with me." He gestured again and Sebastian's eyes rolled up in her head and she slumped to the ground. "She will sleep for now. Come."

Steel stepped around the man and kneeled beside Sebastian. He lifted the sunglasses away from her eyes. They were closed and her chest rose and fell slowly. Steel knocked on the door to the truck and the driver, a bald headed man with several piercings looked through the open window. "Will you take care of her?"

"She is my boss. I will watch over her. Just don't ask me to mess with that medicine man."

The holy man took Steel by the arm. His skin was hot and dry as old leather. Steel was still reeling from the flashback. He had met Arthur Knight before and the two of them never realized it. His heart burned with the memory of Vega's hug and his promise. Had he always made such promises?

"God always surprises me." He whispered as much to himself as to the priest.

"More surprises await you, my son. Your memories of the past come when you need them. They teach you what you have forgotten. Some knowledge is hidden in our memories until experience comes to temper that knowledge into wisdom. Your hatred of your father will be you undoing unless you learn to forgive."

Steel stumbled across the dirt road under the man's direction. "How do you know such things?"

"Your angel told me." The man led Steel across the road into a flat expanse of stone and trees. A domed church sat in the center of the space. "This is the Church of Our Lady Mary of Zion." The man said. "It is the most holy church of the Ethiopian Orthodox Church. I will not allow the unworthy to violate the grounds of this church."

"Why am I so special?"

"Your angel told me you were coming." The man said quietly. "I want to show you something. I am the Guardian Monk of the chapel." He pointed to a square stone building sitting in the back corner of the property. It was beautifully constructed and surrounded by a red, wrought iron fence with arrows pointing outward at the top of each spire. The guardian monk opened a gate in the fence and led Steel up the stairs into a domed entrance. The interior was adorned with murals depicting a woman, most likely Mary, in various moments of her life. But whatever was in the center of the chapel was obscured by a huge, red curtain.

The man pointed to the corner of the room. There, another red curtain stirred in a breeze that seemed to come out of the very wall. "Your journey is only beginning, son. God has chosen you for an important task."

Steel's anger built and his face warmed. "I don't care about a task. I'm here to find the Fallen Throne so I can save my friends."

The man nodded. He placed a gnarled hand on Steel's chest. It was hot and the touch of the hand on his skin brought comfort. Steel's anger abated. Calm came over him.

"You have a good heart. You are a man who has the righteous anger of God at your disposal. Always make sure you unleash that anger on those who deserve it. A great loss is coming and you must be strong in the time of tribulation. You must keep your eyes focused on the task. Sebastian is not to be trusted now, but in the future, you will have to trust her. God is not finished with her yet."

Steel swallowed. "I don't know--"

"You have a great hatred in you, my son. Defeat it or it will destroy those whom you love." He withdrew his hand and Steel gasped as the warmth and comfort receded. The holy man pointed to the curtain.

"You must now pierce the Veil. Another awaits you to show you to your fate." He gestured to the curtain. Steel hurried across the room and paused as he looked at the center of the chapel.

"Is that the Ark?"

The holy man smiled. "There are many things man cannot look upon and live." He moved to the center where the red curtains hung motionless. He reached inside and through the slit Steel saw a wooden chest; flashing gold, an angel's wing. The holy man pulled out a long leather bound object with a sling. He handed it to Steel.

"You will need this."

Steel took the formless canvas bag and felt something thin and flat within. He slung the bag over his back and nodded. "Thank you."

He turned and stepped through the billowing curtain and was some when else. The room was not there and was there. He floated and walked and flew. The air was thick with incense and he was not of this world. A priest stood over a dying lamb. The priest had colorful stones on his chest and incense filled the Holy of Holies and the great purple veil hid all from the world and it tore, serrated from top to bottom obliterating the priest and there on the ceiling was the cross with the God man. His eyes were deeper and more powerful and more loving than was infinitely possible and He looked at Steel as blood ran and he said, "Father forgive them." A drop of blood fell from the crown of thorns and hit Jonathan between the eyes burning like molten gold, eating through his skull to his brain and igniting a fire that consumed his entire body and he was falling through darkness and ashes and light and death and life as the fire coursed through him to his heart and made it beat slowly then quickly and the thing he carried, the sword on his back glowed with holy light and he knew. He knew why he was but not who he was. The who was no longer as important as the why.

Steel stepped through the curtains and out into bright sunlight. He blinked even as the vision or journey or whatever he had just experienced began to fade leaving behind only the conviction. He had to stop Cobalt. He had to save Josh and Cephas and Theo. A holy man similar to the other stepped into view and Steel saw they were at the bottom of a narrow trough carved into the stone.

"Welcome, avenger, to Lalibela."

Steel looked around at stone and dirt walls. No curtains. No chapel. He shook his head. "Where?"

"You are a far distance from your first stop, my friend. You are here in the underground churches of Lalibela."

The priest led him down the long, dusty trough and up into the hot mid morning sun. Steel looked down from ground level at the church carved from the very rock in the shape of a cross. The church was carved out of the ground and its roof was level with where they stood.

"This is a most revered and holy place in Ethiopia." The man said. "The reason Christianity survived here is because of its remoteness. The Islamic conquerors couldn't get through the surrounding mountains to this area. Christianity rooted here and has remained."

Steel looked around at the mountainous terrain covered with green shrubs and short trees. It was more arid than Axum. Worshippers adorned most of the ridges and walkways among the churches. They moved in lines of weaving white robed men and women. From a nearby church deep in the ground, he heard singing and chanting. The air was thick with heat and incense. But, more than anything, he felt the extreme holiness of this place. It permeated the very rock on which he stood.

"And your people carved these churches out of the rock?"

"With hammers and chisels and some of them 800 years ago."

The priest led him around a high ridge and down a rocky ramp into a ravine. Below them stretched an amazing sight. The church had been carved out of the rock and was over one hundred feet in length. Columns adorned the front and a wooden doorway led into darkness. But, the roof was covered with metal.

"The roof collapsed on this church several months ago and killed eighteen church members." The priest led him toward the back of the church. Here, the rubble of the collapsed roof had cascaded down like a land slide and formed a huge pile of broken stone. At the very edge of the collapsed stone walls, the priest pointed out a narrow tunnel carved into the rock. The tunnel led beneath the rubble.

"What you seek is down there in the oldest part of the church. This is the oldest sanctuary in the area. Many of us have never set foot in that part of the church even before the roof collapsed. Those that did are still there beneath the stones. God will be with you, my son." Steel looked up at the surrounding walls of sheer stone from which the church had been carved. Over a dozen of the priests stood looking down upon him. They began to pray quietly.

Steel wiped sweat from his brow and stepped over broken columns and chunks of rock from the shattered church walls. He squeezed beneath an angled column of red rock and stepped into the cool air of the tunnel. Hands touching these walls for centuries had polished the rock to a sheen. How many of them had been here when the ceiling collapsed? The odor of stale decay was barely noticeable. How unstable was this tunnel that families would leave their loved ones to decay rather than risk more death?

An opening roughly ten feet in diameter had been broken through the roof of the tunnel from ground level. Chunks of the roof had shattered and covered the floor with red hued boulders. Steel spied a withered hand protruding from under one of the stones. He stepped over another large broken chunk of rock and his foot slid on the slick surface. He fell backwards onto a tilted slab of rock and slid downward into darkness. He fell into open air and crashed with a thud onto sand. The blow drove the air out of his lungs. He lay there gasping for breath looking up at the crack through which he had fallen. Sand trickled through the opening and the waning sunlight filled the air with dust motes. He slowly stood up and stumbled backwards, struck something hard and sat down with a thud.

He glanced over his shoulder at the hulking shape towering above him. He leaned against rough, dark rock and stood up painfully. Black rock shot through with pores and white streaks sat in the center of the chamber. It was roughly the shape of the Fallen Throne but was far from a shiny crystalline structure.

"It's covered with lava."

He whirled. Vivian Darbonne Ketrick stood at the top of the rock down which he had slid.

"You!"

"You didn't expect Cobalt to come and get his throne did you?"

"No, he sent his witch." Steel spat dust and rock out of his mouth.

"Honey, I'm far from being his witch. I have two jobs. One is to stop the man. The other is to make sure he doesn't kill you." She wore a dark one-piece jumpsuit and she sat on the edge of the sloping rock and slid into the chamber. She came to rest at his feet and held up her right hand. "A little help, please?"

Steel glared at her. "I'm not much of a gentleman right now."

Vivian nodded and stood up. She dusted red dirt off of her clothes and straightened her hair. "Let's get one thing straight. Cobalt took one of my demons hostage and I don't get it back until we return with the throne. Got it?"

"Maybe he should have taken all of them." Steel said.

"Honey, that's happened before and I was a basket case." She stepped around him and studied the throne. "I'm guessing the throne must have ended up in a lava pool somewhere in the past. I'm surprised these Ethiopian priests haven't chipped away the stone."

"They respect the divine, Vivian."

"I do, too, honey child. Now, how are we going to get this little old throne out of here?"

Steel grabbed her by the arm and whirled her around. He shoved his face into hers. "I'm sick and tired of your games, Vivian. You led Cobalt to Josh and Cephas and left Theo to die. I should kill you right here!"

Vivian giggled and then laughed out loud. "Oh, baby, and Cobalt thinks you're so righteous. This is why you can touch me. You're just as full of violence and vengeance as I am." She pulled her arm out of his grasp and frowned. "We are no different, Jonathan. I am a slave to my demons and you are a slave to the angels. Neither one of us wants this life, but it's what we're stuck with until something better comes along. Now, put your famous fury to bed so we can get this throne to Cobalt or you'll never see your precious friends again."

Steel gasped for breath and backed away, wiping sweat and dirt from his face. He fought for control and pressed the red tide of his anger back down into the recesses of his mind. "We're nothing alike." He whispered.

Vivian smiled. "There's my baby." She gestured with her right hand and the rock above them lifted away in huge chunks. Green light bathed the chamber and Steel looked up into the open iris of the same flying craft from which he had fallen. The green light pulled the rock chunks upward and the force threw them aside. "Let's take the Fallen Throne to Cobalt and get this over with."

The throne rocked back and forth beneath the force of the green energy beam and almost toppled over. Steel reached out to protect himself from the toppling throne. The second his hands touched the rough rock, silence fell. He looked around. The world had ceased moving; time itself stood still. Motes of dust hung motionless before him. Vivian was frozen in time.

"You must see something."

He whirled. The priest stood before him. "You must know what the defiler is planning. I give you this knowledge." He gestured with his bony hand and Steel was somewhere else.

Chapter 32

Out of the mist, the light swirled and coalesced into the shape of a man. Kabal pulled his molecules together as he left the extra dimensions behind so that he could become substantial in his new form. As a glowing man shaped mist, he regarded the village in the valley below him. Already, the morning sun was peeking over the far hills and the light pierced his form. What should he look like? Other Messengers had appeared to humans as flying creatures. Kabal, on the other hand, loathed the idea of wings made of feathers. His form should befit his status. After all, he was second to Lucifer himself.

Kabal allowed the brightening sunlight to flow through him even as he settled on his chosen appearance. He assumed the stature of a man with long, silver hair and bronze skin. For his eyes, he chose a shimmering lavender hue with slit pupils. For his wings, he decided to ignore the feathers. They were a sign of weakness. Instead, his wings unfurled behind him covered with small, iridescent scales. As he assumed solid form, the sunlight refracted from his wings and filled the air with rainbows. The humans would be very impressed.

Kabal launched into the air and soared toward the valley. He swooped down through the cool morning air. His goal was the palace in the center of the village. A group of guards reacted as his shadow passed over them. They raised spears in his direction until they realized he was a Messenger. Kabal glided around the huge dome of the palace and landed gently before the guards. Rainbows danced over their armor and they fell to their knees.

"Messenger, we did not know." The lead guard said.

Kabal resisted the urge to sever the man's head with a swipe of his wings. He had to remain civil. "I must speak to your king."

The guard stood shakily and motioned through the huge cedar and brass doors toward the interior of the palace. "I shall escort you, Messenger."

Kabal followed the guards into the palace. He studied the half finished walls and the partially gilded columns. The king was having difficulty completing his palace. Half starved slaves worked in the far shadows carrying huge stones and buckets of mortar. Overseers jabbed at the laggards with heated metal sticks. The air was rich with moaning and screams. Kabal smiled. This was going to be easier than he thought.

The huge, golden doors to the throne room opened before him. Sunlight filtered down from windows in the huge dome and illuminated the raised throne of the king of this land. The throne sat at the edge of a huge, metal disc. Pitch black metal composed the disc and artisans worked around the edge, tapping with diamond tipped rods at the surface. The translucent blue crystal throne glittered in a ray of sunshine. The humans had managed to carve the throne into the shape of a sitting Messenger. Two huge wings towered above the throne and two wrapped around the feet of the throne. The back was the chest of the Messenger and its upper arms were raised above the throne. The lower set of arms rested on the Messenger's legs forming the armrests of the throne. In the place of the Messenger's head a mask with oblique eyes and a slit of a mouth allowed the rich, crimson glow of the Bloodstone to shine through.

"A Messenger requests an audience with the king." Kabal said.

The man on the throne stood. He was tall and well muscled wearing a gold and maroon tunic. His dark hair fell upon his shoulders and the Mark glowed on his forehead. His black eyes simmered with fury. Around his neck, a golden chain dangled. A fiery maroon jewel the size of an egg glittered over his chest. Around his waist, he wore a golden belt of beaten, hinged metal. A scabbard hung from the belt holding the most beautiful thing on Earth. The king's hand strayed to the crystalline sword in the scabbard. At the touch of his hand, the sword glowed with a lavender light and the crystal surface of the throne glowed with it.

"What does the Creator want with me now?" He bellowed.

The artisans scurried away from the disc as the king moved to its center. Kabal was impressed with his arrogance. Such behavior to one of the Messengers of Light would have brought instant retribution from God. But, Kabal's master was more patient with such arrogance. Arrogance was one of his tools.

"I do not come from your deity, my king. The possessor of the Bloodstone deserves much respect. Rather, I come from another who brings you an offer."

The king raised an eyebrow and unfurled his cape as he stalked around Kabal. The smaller Bloodstone at his neck glowed with the energy of his anger. He drew the sword from its scabbard and held it at his side. "Who is this one making an offer through you?"

"You know him as Lucifer, the fairest of the Messengers."

The man paused in front of Kabal. He raised the sword and pointed its gleaming tip at Kabal's chest and laughed. "The fairest of the fallen Messengers. What power does he have?"

Kabal smiled and held up his right hand. Fire danced from it into the air and formed into a globe. "Do you have a servant who deserves punishment?"

The fire reflected in the king's eyes as he studied Kabal's hand. Flames reflected from the Bloodstone and he lowered his sword. The king motioned to the center of the disc.

"I will always have enemies, Messenger." The king motioned to a nearby wall where a dozen men were shackled. He gestured to a guard who ran to the nearest prisoner and brought him forth. The man wore only a tattered loincloth. The king motioned to the center of the disc as he backed away. The guard threw the man into a slightly depressed circle in the center of the sphere.

"My king, I beg for mercy." The man screamed. When he tried to move from the circle, blue streaks of lightning

coruscated over his body. He screamed in agony and pulled back into the circle.

"Only one who wields the sword may stand in the center of the Infinity Disc." The king said. "I was planning on using the Bloodstone on this man, but the effort would have drained me of energy. I am interested in what you have planned for him." His hand caressed the Bloodstone as he looked at Kabal.

"You have a larger version, do you not?" Kabal asked.

The king's eyes flickered with hesitation and then he pointed to the throne "The main stone sits at the head of my throne as the Messenger can see. You know it well. And, I carry a fragment of it around my neck to augment my powers. This you know."

"Your god put the Bloodstone at the head of your throne for a reason." Kabal leaned closer and whispered into the King's ear. "It is there to remind you of the innocent blood you shed. That is why it is red. That is why it is known as the Bloodstone."

The king's eyes filled with anger and a hand strayed to the Mark on his forehead. "Do not remind me of my past failures, Messenger. Failure is but a stepping stone to power. Now, do with this creature as you will."

Kabal straightened. "As you wish." He tossed the ball of fire onto the man's head. In an instant, the ball expanded and surrounded the man with greenish fire. The man began to shriek even as his skin sizzled and blistered. The ball contracted in upon him and pushed his struggling limbs down until he was a compressed ball of blackening flesh. With a sudden gush of heat, the ball of fire exploded and released smoke and the rancid smell of burned flesh. Ashes settled onto the disc's surface.

"Impressive." The king stepped back from the pile of glowing ashes that had once been a man. He flicked his wrist at one of the nearest artisans. The man ran forward with a brush and began to clean away the ash. "But, it is a feeble parlor trick compared to the power of the Father of Creation."

Kabal placed his hands behind his back and gathered his wings close to him. He paced around the king. "That is true. But, has not your deity promised you a land of your own?"

The king followed Kabal with his calculating gaze. "I have a kingdom."

"A small and paltry land of cowards and outcasts from your earthly father's kingdom. Look around you at this half finished palace. Your subjects are weak and sickly. Your father's kingdom in the far valleys flourishes while yours hovers on the brink of extinction. How many kingdoms have you established? How long have you and your followers wandered about this land? Does not your father's kingdom cast a shadow upon yours? You are doomed to forever fail. Tell me, how long have you ruled here?"

The king touched a ringed finger to his chin. "I am seven hundred and thirty-two years old. But, I have only ruled these lands for a scant thirty years."

"And, for most of those centuries, you have born the Mark." Kabal paused and glared at him.

"It is the sign of my curse."

"It is also a sign of the covenant your deity made with you. No one may kill you. That makes you very powerful. Is not that the real reason you have been allowed to possess the Bloodstone? To protect yourself from those who desire your throne?" Kabal leaned closer. "And, your deity has promised you your own land. I am here to help you claim it."

The king tilted his head in thought. "What power does a fallen Messenger have to offer a man such as I a new kingdom?"

Kabal had been waiting for this moment. He clapped his hands and motioned to the dome above him. With a boom, the dome shattered near its center and the falling glass and rock swirled away toward the far periphery of the throne room as the Master descended. Light surrounded the Master as he settled slowly toward the throne. The guards hurried to protect the king. But, the descending figure, surrounded by his magnificent light settled onto the throne. Such a position of power should have been Kabal's.

Lucifer arose from the throne. He was indeed magnificent. His golden hair shimmered in the sunlight. His breastplate glittered with diamonds and gold. His huge, white wings unfurled behind him and Kabal felt jealousy at their glowing whiteness; the substance of pure light.

"My good king, I come to offer you something you could never have imagined." Lucifer smiled.

To his credit, the king never flinched. "Why should I consort with such as you? You are but the prince of the fallen, Lucifer."

Lucifer glanced at Kabal and shrugged. "Your own god has cursed you. He gives you a supernatural stone that can kill with a mere thought but at the same time robs you of your vigor. And, although you have sired children have you not longed for a suitable queen to sit beside you?"

The king tensed and drew a deep breath. "I was told that one day I would find my true wife."

"And, now is the time, my king. For your wife resides in a far land. All this while, you have waited for what he has promised you -- a far land and this is what you settle for?" Lucifer motioned to the shattered dome above him. "Is not your god the god of the universe? Has he not created the stars and the planets of the heavens? Are there not worlds uncountable as numerous as the grains of sand on the shore? I know for a fact there is another world beyond this one."

The king's eyes squinted in confusion. "Another world?"

Lucifer pointed to the sky as blue as a robin's egg just beyond the ragged hole in the dome. "Out there are worlds beyond this one. Your god has created you a world of your own. But, in your limited thinking, you did not claim it. I can help you claim it."

Lucifer snapped his fingers and the sky darkened. Shimmering stars appeared in the darkness. He motioned with his hand and the stars seemed to fall toward them. The king's guards fell back in fear. But, the king did not move. He merely clutched the Bloodstone in his fist and it glowed feebly with killing energies. Kabal marveled at the interest that filled the man's eyes. His curiosity would prove to be the beginning of his doom.

Lucifer extended his wings and stirred the air. The stars fell like gleaming points of light and whirled in the wind. One star settled into Lucifer's outstretched hand. He breathed on it and it expanded. The point of light became a cloud of iridescent beauty expanding to fill the entire throne room. In the interior of the cloud, more stars gleamed and one star in particular enlarged into a ball of flame. Lucifer nudged it aside and pointed to a tiny green and blue dot. He held it in his hand and it expanded into a globe. Kabal watched the king's eyes fill with fascination as the image of the world reflected in his corneas. Clouds swirled over azure oceans. Green and brown crossed continents. Lucifer held up the image to the king.

"Imagine your own world complete with lands and seas and cities. Imagine being the undisputed master of such a world. Imagine every subject falling at your feet in abject praise."

The king stepped back from Lucifer and the fascination in his eyes was replaced with irritation. "Why wasn't I told of this?"

"You never asked." Lucifer laughed. "If you never knew such a possibility existed then you would never make the right choice. Poof." Lucifer closed his hand and the planet, sun, and cloud disappeared leaving them in darkness. "It never existed. If you want it to be real and if you want to have it, all you have to do is ask."

The king blinked. "And, why have you bothered to tell me this?"

Lucifer glanced at Kabal. "Let's just say, I would like to go with you. Why, all of my Messengers would be at your beck and call. With the Mark on your forehead and the Bloodstone at your beck and call, no one can kill you. You will live forever and we will be your servants."

The king's eyes glittered with the power of his Choice. Kabal saw it coming. The king was walking down the path to his own doom. How clever the Master was. But, Kabal was cleverer. He unfurled his wings and the rainbows danced around him.

"Master, allow me to show him our army." Kabal said.

Before Lucifer could reply, the darkened sky above the hole in the dome filled with singing. The king's face turned up as a legion of the fallen descended from the sky. Their voices thundered with praise for their Master. Lucifer hesitated and glanced once at Kabal. He had not known of Kabal's plans. He had not considered this thunderous chorus. Would that be enough to hide Kabal's true intentions? He hoped so.

"Here, my king, is the army we have assembled for you." Kabal proclaimed.

Messengers thronged the throne room, descending in dozens to land on the stone floor. The suspicion on Lucifer's face melted away beneath the sea of praise. He turned and drank in the assembled fallen Messengers and their adoration for him. He gleamed at Kabal.

"I asked Kabal to assemble an army. He has done well, my king. Silence!" Lucifer shouted. The assembled Messengers fell silent and kept their gaze on Lucifer.

Kabal was proud of himself. The master did not know the army was loyal to Kabal. Each Messenger had assumed a pleasing appearance per Kabal's instructions. Kabal felt the weight of their stares on him, reeled under the sheer majesty of their collective power. He bowed to Lucifer and then addressed the king.

"Sir, your army awaits your decision."

The king looked at Lucifer and then turned slowly as he studied the assembled fallen. He paused. "I want my world." He shouted.

Kabal sensed the greasy, shiny unction of the thing before it appeared. Above them, the Messenger of Light descended into the throne room. Kabal cast covert glances at his army. Don't react, he thought. Don't let the king see your true personality, he said beneath his breath. Even so, the army stepped back as the Messenger of Light alighted on the stone floor in front of the king. Above them, the sky darkened and thunder echoed in the vast hallway. He was resplendent in a dark green tunic and his wings were indeed covered with feathers. Feathers! His blond hair was tightly curled against his head and his green eyes glittered as he stood before Lucifer.

"What manner of deception are you attempting, Father of Lies?" His voice boomed across the chamber.

Lucifer raised an eyebrow and laughed. "Deception? I am

merely reminding this good man of his choices. You and your Master have carefully hidden the world he can have. I would say you are the deceiver here. Not I."

The king stepped between them. "Who are you?"

"I am Donnatto, a servant of the Most High God, Creator of the Universe. You should not let Lucifer's words deceive you."

The king glanced once at Kabal and back at Donnatto. "Is it true? Is there a world for me?"

Donnatto's gaze broke from Lucifer and he stared down at the king. "Yes. You were not told because it was not time."

The king's eyes danced with anger. One hand clutched the Bloodstone and the other moved to his forehead. He touched the Mark. "I have lived here in this wretched palace and could have had my own world?" He paced before Donnatto and his fists clenched in anger. "The Father of Creation promised me my own land with a wife who would be my queen. I am tired of waiting. I want it. Now. You will deliver me and my subjects," he paused and swept his hand around the room, "and my army to my new world."

Donnatto sighed. "You do not know what you ask."

The king pressed his face close to the Messenger's rigid features. "I do not know what I ask? I want my land to rule and I want it now. If it is promised to me, then I want it."

Donnatto closed his eyes in defeat and shook his head. "Then, you shall have what you deserve. Place the sword in the hands of the throne."

The king strode across the disc and stepped up into the seat of the throne. He could barely reach the closed fists of the upraised hands of the image of the messenger. The hands came to life and took the hilt of the glowing sword and raised it high above the throne. The throne began to glow with life.

The king stepped away in amazement as the Bloodstone behind the mask came to life. The mask changed into the face of an angry Messenger. "Who brings the Fallen Throne to life?" A booming voice echoed across the room.

"I do." Donnatto said quietly. "I request you open the portal."

"Where is the sacrifice?" The Fallen Throne asked as its glowing eyes of fire turned toward Donnatto.

"There will be none on authority of the Father."

The Messenger's gaze shifted until its eyes came to rest on the central depression of the Infinity Disc. It brought down the sword swiftly, stabbing at the air above the seat of the throne.

"No blood was shed by order of the Father." It said and red beams of energy shot out of the eyes and into the central depression of the disc. It glowed with ruddy, crimson heat and the entire disc shimmered with energy and moved like the surface of a pond. A mirrored sphere expanded out away from the center and stopped just beyond the periphery of the disc. It pulsed in and out with a crimson glow and the mirrored surface cleared allowing a view of the other world. Distant mountains towered over a lush, green meadow.

"What is it?" The king walked around the sphere. "As I move, I can see more of this world!"

"The Portal is open." The Fallen Throne said. "Let any who enters beware."

"Your world, my king." Donnatto said. "But I must warn you. Lucifer has asked God for your soul. He has asked to test you. This is your test. Listen to the words of your God and do not ask for this. Refuse this world now and save your own soul."

Lucifer pushed the king aside before he could respond. "Do not listen to this one's lies, my king. He has already admitted to deceiving you. Let me be the first to cross over into your new world, my king. For once I arrive, then all of these meddlesome Messengers of the Light will fall before me."

"You will not pass!" The Fallen Throne shouted. The arms flew up into the air and it tossed the sword over the sphere and into the hands of Donnatto. He slashed the air between Lucifer and the Portal with shimmering light. "No, Lucifer. You are banished from this world."

The king glanced once at Lucifer and then at Kabal as he stepped into the energy of the portal. His image shimmered with the pulsation of the portal's outer edge but he now stood in the midst of the meadow. He whirled and gazed at the sky and mountains around him.

Donnatto held out his arms to block Lucifer's way and from every corner of the kingdom, humans streamed through the air, sucked into the glittering sphere of energy like leaves blown on a mighty wind. Their frightened faces were a blur as they surged through the air into a steady stream of flesh that slid gracefully through the energies of the Portal. Animals followed, bellowing and bleating in a whirlwind of fur and flesh. The air was alive with life swirling and whirling into the Portal.

Kabal cast one look at Lucifer and noted that his eyes were locked on Donnatto. Kabal gestured to the nearest fallen and the plan began. With one huge swoop, the fallen army of Kabal joined the stream of humanity through the Portal. Donnatto turned even as Lucifer did and screamed in protest. Lucifer reached toward the portal and Donnatto launched himself into the air. His huge wings unfurled and he crashed into Lucifer. This was exactly what Kabal had planned. He watched as fallen messenger after fallen messenger all loyal to him streamed through the Portal with the writhing, frightened humans. Lucifer and Donnatto spun upward in the air, locked in combat. Kabal laughed and made his way to the Portal as the last of his army disappeared along with the last human. He cast one glance up at the two battling Messengers and turned to the Portal. Just inside the interface with this reality, a tall Messenger stood. His hair was pale and his eyes an intense blue.

"Gabriel?" Kabal hissed.

"You may not pass, Kabal. Now that your minions have tainted this world, know that they will not be unopposed." From above Gabriel, white robed Messengers poured through the portal from the heavenly dimension. "You are forbidden." Gabriel said and stepped through the interface into this reality.

"No!" Kabal screamed.

Gabriel smiled and gestured to the Fallen Throne. The portal collapsed in upon itself and Gabriel disappeared.

"None may use the Portal again without the ultimate sacrifice." The Fallen Throne said. Its hands reached down and

grabbed the Bloodstone and hurled it upward into the open air of the throne room. It flew upward through the open dome and out into the cloudy darkness. From around the throne room, the collapsed stone of the dome vibrated and shot up into the air. Piece after piece shot forward and adhered to the surface of the throne until it was covered in rough, black stone. Then, the throne spun upward into the air and out through the hole in the dome as thunder crashed and lightning split the darkness.

The Infinity Disc tilted up onto an edge and spun faster and faster until it was one spherical blur. It disappeared into thin air with a loud pop.

Kabal fell to his knees and shrieked in despair. Above him, Donnatto and Lucifer continued to fight until Donnatto pulled away and looked down upon the empty throne room. Donnatto shoved Lucifer away and flew upward through the shattered dome after the vanishing throne. At his passing, the clouds dissipated and the darkness receded to reveal a bright, morning sky.

Kabal stood shakily to his feet as he realized his plan was now destroyed. All that remained on the stone floor was the glittering crystalline sword. The shadow fell upon him and turned to see Lucifer settle on the stone in front of him. Lucifer's eyes burned with a fury that sucked the very breath out of Kabal. The power of his Master's hatred shoved him to the floor. The heat of his anger burned away Kabal's fair hair and the beautiful clothes. His wings shattered like brittle glass. He was left small and shrunken on the stone of the abandoned palace even as Lucifer towered over him.

"You dare to try and thwart my plans, Kabal?" Lucifer roared. "Do you realize what this has cost me? I had my own world in the palm of my hands and you threw it all away!"

Kabal groveled in pain on the stone. The heat and hatred of his master was consuming him from the inside. "Please, have mercy on me. I am the shining star of your eye. I am your greatest companion. I gathered the army you led against Father Creator. Have mercy on me."

Lucifer breathed deeply and the heat began to lessen. "From this day forward, your loyalty to me will be unquestionable,

Kabal. From this day forward, you will serve me hand and foot and you will never turn against me. Do you understand?"

Kabal stumbled weakly to his knees. His burned and crisped flesh peeled away as he clasped his hands and bowed his head. "I swear an oath of undying fidelity to you my one and only Master." He wheezed.

Lucifer nodded and launched himself upward into a bright and cloudless sky. Kabal stood shakily and shook the blackened scales from his form. He glittered like black onyx. He was a being stripped of hair and flesh and wings. He turned to the empty throne room. He glanced down at residual shards of the maroon stone. They were scattered across the floor.

"What has happened here?"

A group of confused humans appeared from the corridor. They were dirty and wore torn clothes. The man leading the pack of humans had long, matted black hair and his eyes were a muddled image of the king's.

"I am the king's son, once heir to his throne. We heard the battle and the shackles fell away from us in the dungeon. What has happened here?"

"Your father has taken his kingdom and has left this world." Kabal snarled.

The young man's eyes filled with fear. "Who are you?"

"I am Kabal. I tried to stop him, but he would not listen and my Master has punished me for this." Kabal stepped closer. "It would seem that all who did not pledge allegiance to your father were left behind. The kingdom is now yours."

The young man blinked and glanced back at those who followed him. His eyes fixed on the crystalline sword at his feet. Slowly, he stooped to pick it up. He held it up to the light coming in from the broken dome. It caught the sunlight from outside but did not glow with the power of a king's touch. "I do not know how to be a king."

Kabal smiled and his shiny ebon skin caught the flickering light of the torches carried by the remaining humans. Perhaps he could yet save something from this disastrous day.

"If you will serve me, I will do everything I can to find a way to bring back your father's people to their rightful ruler."

Kabal stepped closer to the man and placed one dark talon on the man's shoulder. His eyes filled with fear.

"I don't believe I have a choice."

"Good." Kabal motioned to the crystalline sword in the man's hand. "The sword is yours to show the world you are the new king. You must rebuild the kingdom. And, when I find your father's throne and the Infinity Disc, I will come back to you. And on that day, I will reopen the portal and claim my world. Not Lucifer's. For now, use this as your father did to consolidate your power." He handed a shard of the Bloodstone to the king's son. He cast one last defiant look at the heavens and launched himself into the sky.

Part 6

E. B. E. (EXTRATERRESTRIAL BIOLOGICAL ENTITY)

Those who accept the authority of the Bible and embrace a Christian worldview take different positions on whether God might have created intelligent life on other planets. This question has been debated at least since Thomas Aquinas discussed it nine centuries ago.

Scholars who believe extraterrestrial intelligence (ETI) physically exists see it as a display of God's creativity and power. They argue that a God who so obviously enjoys creating, a God of unimaginable power, should not be expected to limit His creative expression to just one planet and its one species of spiritual beings.

Hugh Ross

Chapter 33

Steel blinked and was back in the chamber. What had he just seen? Was this real or was it a deception? Did the tenth demon really think there was another world over which he could rule? The priest was gone and time resumed its movement.

The throne and Vivian floated upward and Steel's feet left the ground. They moved through the opening above and hovered beneath the shadow of one of Cobalt's ships. Below him, the vast expanse of Lalibela's underground churches spread out in a stunning vista as he emerged from the sunken ruins of the church. Children and adults moved out of houses and huts and pointed to the ship. The throne disappeared inside and he followed. The priest who had met him outside the church walked into view, arising from the dust and debris of the collapsed church to stand atop a crumbling wall. "Go with God, my son." He shouted. "And protect the sword."

The iris closed with a hiss. The buzzing noise shocked him into motion. He whirled to see a nightfly standing between him and Vivian. The nightfly held the inertia gun in one of its four hands.

"Reyjacklik, there is no need to point that gun at Mr.

Steel." Vivian said. "He will cooperate if he wants to see his friends again."

The nightfly lowered the gun. Steel tried to focus on the creature but it seemed to vibrate in and out of focus. "He is of the Light. He deceived me before." The thing said in a high pitched warbling voice.

"I know." Vivian motioned to Steel. "We can go to the lounge while Rey baby flies the ship."

Steel allowed the canvas bag with the sword to swing nonchalantly behind his back. He hoped she wouldn't notice it. He stepped around the creature and it flinched when he leaned toward it. It did not like Steel's 'Light'. He followed Vivian up the same ramp he and Arthur Knight had descended the day before. "Where is the rest of your crew?"

"Cobalt only trusts me with one of his precious creatures. Rey has pledged his loyalty to me." They passed a white and lavender panel in the corridor. It was jarringly out of place in the smooth, curved contours of the ship.

"What's this?"

Vivian paused. "I guess I should acquaint you with our escape pods. Cobalt added them just before I came to get you. Seems our final stop may place the occupants of this ship at jeopardy. That's the very word Cobalt used and then he told me another one of his endless stories about the Titanic and its lack of lifeboats. There are four escape pods, one at each quarter of the turn. Just hop inside, buckle in and press the big green button. It's programmed to land in Area 613."

"Area 613?"

"Cobalt's compound. It used to belong to the government until he bought them out. I wish I had his money back. What I could do with that cash." Vivian motioned further down the corridor.

"You are already filthy rich." Steel followed her.

"The more you have, the more you want, honey." She stopped outside an open door and gestured inside. A window on the far side showed passing clouds. Couches and tables made the space seem almost relaxing. "Would you like something to drink? Maybe a martini?"

Vivian sauntered over to a wet bar. Steel watched the clouds speeding past them and he slid the canvas bag off his shoulder and hid it behind a puffy chair. "You are being far too kind. Something about you is different. I'm betting Cobalt took more than just one of your demons."

Vivian dropped a glass and it shattered in the sink. She leaned against the edge of the counter and drew a deep breath. "How did you know?"

"You're weak, aren't you? You've been acting almost civilized. Showing me around the ship like a flight attendant. Without your demons, the mighty Vivian Darbonne without an apostrophe is just an ordinary human."

Vivian glared at him with upswept eyes. She drew a deep breath and reached for another glass. She poured it half full with an amber fluid from a bottle. With one quick throw, she downed the liquid and wiped her lips with her free hand. "He took all of them. Temporarily."

"How does it feel, Vivian? To be normal?"

"It sucks!" She studied her reflection in the glass. "How do you do it, Jonathan? How do you face every moment with nothing at your beck and call but your wits?"

"I do it with the help of God." Steel moved closer.

Vivian laughed and poured another drink. "So, you are no different from me. You have your power, I have mine."

"Only my power won't consume me and take over my will."

Vivian downed another swallow and glared at him. "Not my will but thine be done. Isn't that what the Savior said? Whose will controls you, Jonathan? Have you stopped to ask that question? Since the day we met, you've been a puppet on divine strings moving to the jerks of an unseen power. You're not your own, honey. You belong to a master just as I do. At least I have the honesty to admit I'm empty right now. I know who controls my tomorrow and when we get to the Eagle's Nest, I'll have my power back. Only, I will control my demons." She placed the glass in the sink and laughed. "While you will keep wondering when you'll be able to control your own fate. You belong to Him, Jonathan. You will never be free."

Steel clenched his fists and the anger stoked within. Was she right? What had the monk said? He had found his purpose? If so, where was the peace? The acceptance? All he wanted was to find Josh and Cephas and destroy Anthony Cobalt.

Vivian giggled and stepped around the wet bar. She reached out and touched his chest with a finger. "See, I can touch you now without getting burned."

Her hand flattened against his chest and Steel tried to pull away but she stayed close to him. She slumped against him and pressed her pliable body against his. "See, you are so angry right now. You probably want to strangle me, don't you baby? It's your rebellion that makes you weak, Jonathan. Cuts you off from your power. You want to lash out. But, at the end of the day; at the moment of Choice, your Master is controlling you. Right? You have to do the right thing? And, killing me is wrong." She ran a finger along his lower lip. "So wrong. Just like this." She lifted up and kissed him on the lips.

Steel collapsed away from her, stumbled over the couch and fell to the floor. His breath came quick and hard and he fought for control. "Is that how you seduced the deputy?" He whispered. "Right before you killed him?"

Vivian tensed and tilted her face back and forth as her skin reddened. Her eyes widened and she screamed. "Don't you dare!"

Steel stood up. "I saw his body. He looked a lot like me, Vivian. Did he have turquoise eyes? Did you look into them as you drove those tripod legs through his heart?"

Vivian launched herself over the couch and grabbed him around the chest driving them back over the coffee table and into the window. She clawed at his chest and then went for his eyes. "I'll rip those eyes of yours out of their sockets!"

Steel grabbed her arms and spun around, pinning her against the window. "You're weak now, Vivian. So let me get one thing straight for you. I don't love you. I don't like you. I hate you with every fiber of my being. I can't wait for your demons to drag your soul to hell! Do you hear me? Whatever it is you feel for me, rest assured I don't feel anything but revulsion for you." He released her and she slid down to the floor, sobbing. "At the end of the day, Vivian, I pity you." His heart pounded and he touched his lips. They were bleeding.

It was then that Steel looked up and gasped. The ship had cleared the clouds and the view outside show a stunning visage of the earth arching beneath them. Against the far horizon, something golden and shiny hung like an ornament in the sky. Closer than the object and nearing them with unbelievable speed was a space station shaped like a nest, all glittering gold and shining glass. They were in space.

Chapter 34

Arthur Knight waited patiently on the bench just inside the mall entrance. He wore a simple black tee shirt and jeans with a dark cap pulled down to hood his eyes. After avoiding Major Robert Miller for years, he couldn't believe he was waiting to reveal himself to the man. Nor could he believe he had let Jonathan Steel talk him into this meeting.

Steel and Cassie had headed for Ethiopia on one of Renee's transports and that left the job of contacting Robert Miller to them. Arthur refused to let Renee meet with Miller so here he sat. A group of children from a day care swarmed around a nearby cluster of electronic rides. This is why he had chosen the mall in the middle of the day.

"You're pretty smart for a dead man."

Knight looked over his shoulder. Miller stood behind him and cold metal touched Knight's neck. "You wouldn't shoot a man in cold blood in front of a dozen little kids, would you?"

"Were you Renee's mercenary?"

"Yes."

"In the hospital?"

"Yes."

The pistol pulled away from his neck and Miller moved around from behind the bench to sit beside Knight. He wore a black hoodie and tucked the gun into the front pocket.

"It's August in Florida and you think a hoodie is inconspicuous?"

"You asked for a truce. I'm here." Miller said keeping his gaze forward.

"Anthony Cobalt has the children." Knight studied the

man's features. A lone twitch in his cheek was the only reaction. Miller drew a deep breath and nodded.

"Of course. He started the project back when you were working with Lucas Malson. He's been the power behind the project all this time."

"And, the puppeteer pulling the Captain's strings."

"Why are we meeting?" Miller asked.

Knight pulled the folded piece of paper out of his pocket. "FBI Special Agent Franklin Ross has been conducting surveillance on a church in Nevada."

"Why?" Miller snorted. "Cobalt wants to have Sunday School?"

"He thought it was a front for money laundering by a wealthy individual. Guess who showed up at the site this morning with the children?"

Miller looked at him for the first time and took the paper. "Cobalt?"

"His church on his property. Ever heard of Area 613?"

"Yes. The secret part of Area 51. It's private property now. Owned by Cobalt." Miller studied the paper.

"That's the official report from Ross to his superiors. They don't want to do anything yet. The FBI has bigger fish to fry." Knight said. "Ross' superiors think the children belong to a family on the church compound."

Miller drew a deep breath. "If they only knew what the children were."

Knight tapped the paper. "Miller, Ross wants your help to break into the compound and rescue not only the children but two civilians, Josh Knight and Dr. Cephas Lawrence."

"Josh Knight?" Miller glanced at him. "Your son?"

"You would know." Knight glanced at him. "It's why I faked my death. If you found out I was the Guardian for the children, you would have threatened him to get your way."

Miller's eyes shifted and he looked far away. "Back then, I would have. The Captain can be very persuasive. But, after the ski slope incident, my motivations changed. Vega is my daughter."

Knight took off his cap and ran a hand through his short

hair. It was time to tell the truth. Finally. "Actually Robert, she's my daughter."

Miller froze and every muscle in his body tensed. His hand slid into the hoodie pocket. He stood up slowly and stepped in front of Knight. He squatted down until his intensely burning gaze burrowed into Knight's face. "Take off those sunglasses."

Knight slid the glasses from his face. Miller stared at him, a fine tremor in his cheek. He finally blinked. "I knew she had an affair. But, with the likes of you? A nerd scientist? Do you know how long I've wanted to kill that man? How long I've suspected he fathered Vega?"

"It's a long story, Robert. I was under the influence."

"Of what?"

"Not what. Who. Lucas Malson. The devil in disguise." Knight tried to look away. But, confession is best delivered eye to eye if it is real. "I've paid for my sins, Miller. And, so has Renee. There's no reason Vega should suffer, too. She needs our help. We can't let Cobalt carry out whatever plans he has for those children."

Miller's breathing slowed and he stood up. His hand stayed in the pocket. He threw the hood back with his free hand and his breathing slowed. He slumped onto the bench again. "I've replayed that time of my life over and over. I wasn't at home when Renee needed me. I was off chasing immortality and power promised by the Maxsapien Project."

"And, as you said, the Captain can be persuasive."

Miller nodded and finally took his hand out of the pocket. "Why come to me? Doesn't Ross have FBI resources?"

Knight pulled out his phone and sifted through pictures. He found the recorded feed from his drone sent to his cloud storage. "On the night of the storm, I met Vega in the swamps around the school and this flying craft showed up. At first, I thought it was yours. But, Robert, this thing is beyond your technology. I had a drone that caught an image of it just before its occupant destroyed the drone. The drone downloaded it to my cloud account before something ripped it out of the sky."

He held up the phone and touched the play button. Miller's gaze shifted to the phone as he watched the video.

"What is that?"

"I think it's Cobalt's ship. Some kind of, and I can't believe I'm about to say this, flying saucer."

"I've seen this before. The XTB-5548. I know it. Part of the retro fitting program at Area 51. It's made from the ship from Roswell. The Air Force task force could never get the thing to work." Miller handed the phone back.

"You're not surprised?"

"No. Cobalt bought a dozen mothballed programs from the Air Force in exchange for his Sunstone technology."

"What's that?"

"Crystallized energy. The future, Knight. A pill the size of an almond will power your car for a week. If Cobalt got the ship to work, then this is bad."

"And, he has a UFO church behind him. The most dangerous people on the planet. Crazy fanatics." Knight said. He couldn't believe Miller had just casually confirmed the presence of extra-terrestrial life.

"Ross needs me, doesn't he?" Miller whispered.

"You've worked with Cobalt. You know experimental aviation. You know this vehicle."

With a sudden explosion of giggles and laughter, the group of children ran around their bench headed for the outside. A woman followed them. "Sorry. They're excited. We're going for ice cream." She hurried after the children.

Knight looked into Miller's eyes. "Besides, Vega needs you. The sooner you get your men and equipment in place around Diablo Boca, the sooner we get the children back."

Miller raised an eyebrow. "Diablo Boca?"

"The Devil's Mouth. A crater located on the edge of Area 613. Cobalt didn't have to build a wall around his church compound. Nature did it for him."

Miller stood up. "Well, Arthur Knight, the Guardian of the Children of the Bloodstone, how do you fight a flying saucer?"

Chapter 35

Josh shrugged inside the white jumpsuit. It itched. He stood in the front row of worshippers in the great stage area. Over two hundred men and women filled the auditorium waiting for the arrival of Magan Celeste. He turned and tried to search the crowd for Cephas but the huge, burly man beside him grabbed his arm in an iron grip.

"Brother, keep your gaze fixed on the stage." He whispered.

The big man had been Josh's constant companion for hours after being taken from Cephas and the message was clear. Behave or Big Foot would cave in his skull.

"Bro, I'm looking for my uncle." He said.

A huge, meaty hand clamped over his mouth and nose. Big Foot leaned down and whispered hoarsely into his ear. "Want to breathe? Keep your mouth shut or I will shut it."

He released Josh's mouth and Josh gasped for breath. He blinked away the dizziness and kept his comments to himself. Magan Celeste appeared on the stage in a puff of white smoke and the worshippers gasped in awe. Smoke and mirrors, he thought. Just illusion.

"Children of the Enochians, welcome to your ascension. Today, you will see the fulfillment of thousands of years of prophecy. But, before we depart this world for the Promised World, you must know that a force of heathen, pagan non-believers waits outside our compound. But, do not despair. Our leader, Anthony Cobalt, the exalted Father of the New World will dispatch these blasphemers in short time."

Josh's heart raced. Jonathan was here? He nudged Big Foot. The huge man glared down at him. "What?"

"You're about to get your gigantic butt kicked. Brother."

Franklin Ross squirmed in his seat. "What is this thing again?"

Miller studied the tactical screen on the dashboard before him. "Northup Grumman Military Assault Vehicle. Modified for my purposes, of course. It can sit up to eight but I've filled the back with weaponry."

"It's hot in here." Ross tugged at the tight collar of his desert colored fatigues.

"Welcome to desert combat, Ross. I told you to stay in your air conditioned hotel room." Miller touched his right ear. "Raptor 3, report."

"This is a sanctioned FBI joint operation. I had to be here." He squinted into the bright sunlight streaming in through the windshield. He had forgotten his sunglasses. "I don't suppose I can smoke in here."

"Messes with the electronics." A soldier standing up behind him said. He manned a swivel mounted gun on top the MAV.

"And, the sand doesn't?"

"Nope, sir. It doesn't." He said.

Ross glanced at the other vehicles spaced out ahead of them positioned at the base of the tall escarpment leading up to Cobalt's Compound. "We don't seem to be a very big army."

Miller nodded as he listened through his earpiece. "Ross I am prepared. We have six V150 Commando Light Armored Vehicles each with one 20mm Vulcan 1300 rounds ready, one 12.7mm MG and 40mm Grenade launcher, one 76mm 8 round ready, 31 in the hull, one 90mm 8rounds ready 31 in the hull, one 81mm mortar with 62 rounds in the hull, and seven TOW missiles."

Ross sighed. "I love it when you talk military."

"Cobalt has one flying craft."

"That we know of. And, it's from another planet. Who knows what it is capable of." Ross said.

"I'm familiar with the craft. It's not that big."

"But, maybe it carries a big stick. An extra-terrestrial stick."

Ross' cell phone warbled. "Hello, go for Ross."

"FBI Special Agent Ross, this is Anthony Cobalt. Could you put me on speaker please?"

Mama Celeste pointed to the back of the stage area. "Worshippers, Children of the Enochians, turn your attention to Diablo Boca."

The crowd stood and moved to the far end of the auditorium. Curtains pulled aside to reveal a huge window looking out upon the compound. From here, one could see beyond the open assembly area with its surrounding buildings to the center of the huge crater.

"Behold the celestial light." Mama Celeste shouted. In the crater below, the white panels of the radio telescope began to move and pivot, turning around to reveal highly polished mirrors. The mirrors moved with a ripple across the vast face of the huge telescope. Worshippers gasped in amazement. Josh tried to pull away from Big Foot but his hand snared Josh's arm.

"I will not miss this miracle, worm." He hissed and shoved Josh to his knees.

The mirrors clicked into place and light from the sun focused on the crystalline sphere above the crater. It began to pulse with yellow light. Within the sphere, points of light coalesced from the glow and tumbled down the six arms holding the sphere above the mirrored surface.

An iris opened on the platform below them and Anthony Cobalt ascended into view. He wore a shining white one-piece jumpsuit with a silver cape.

"Dude, you've got to be kidding me." Josh said.

The window before them slid down into the floor of the

stage and hot, dry air poured over the worshippers. Cobalt stood on a triangular disc that levitated into the air.

"Children of the Enochians, behold the transformation of energy into matter." He motioned to the sphere. "I am harnessing the energy of the sun into millions of Sunstones to power the vehicle of your transformation."

Josh slowly stood up and through portholes along the support beams he watched Sunstones tumble down to ground level, slowly filling the struts. Cobalt now hovered at the same level as the worshippers. He motioned to the ground below. A long slit opened and the children appeared. They wore similar jumpsuits but without the capes. Like Cobalt, each of them stood on a triangular platform as it levitated into the air.

"Behold the Children of Anak." Cobalt's voice rang through the air. The hot wind tossed his silver cape behind him. The thirteen children hovered before the platform and behind them, the support beams filled with Sunstones.

"And, now, fair Enochians, you may ask how we will journey to the point of transformation. Behold my chariot of fire." He spun in the air and motioned toward the crater. Along the far edge of the crater, two sections of the mirrored surface trembled as they slid aside revealing a huge opening on the inside of the crater. Josh pushed his way through the worshippers and Big Foot tried to keep up with him. He studied the black hole in the side of the crater. The opening arched across the rock face in the shape of a ragged smile. Rocky spines gave the impression of sharp teeth. Across the massive crater a wave of pure evil rolled through the air, crashed upon the edge of the near crater rim and flooded the compound. Vega whirled on her levitating triangle and her hands went to her face. She senses it, too, Josh thought. The wave washed over him and he stumbled back, falling among the legs of the worshippers. The sensation of evil was overwhelming, suffocating like black wings wrapping around him.

"No!" He whispered. He looked around. The Children of the Enochians lifted their hands in rapture and moaned with pleasure. Big Foot ignored Josh as he climbed slowly to his feet and pushed once again to the forefront of the group. His eyes were drawn to the wicked grin in the face of the crater.

"Diablo Boca." Josh said. "Out of the mouth of Satan comes evil."

It broke through the darkness and slid effortlessly through the air. The ship had to be at least one football field length in diameter. Roughly disc shaped, it's skin wore the color of a bruise and its periphery bristled with odd protuberances and turrets. The obscene thing looked like melted purple wax imbedded with metal and glass technology; a child's nightmare brought to life in flesh and metal; a ship prepared for war.

The ship paused beneath the sphere and the dome on its apex opened like a hungry mouth. Sunstones streamed from the sphere and into the ship. The dome snapped closed and the ship drew nearer. Josh could hardly breathe in the face of its unspeakable evil. The Children of Anak were on their knees on their individual platforms. Most were crying. Others hid their faces. Vega alone stood erect facing the approaching ship.

Cobalt spun in ecstasy on his floating platform, arms upraised, cape flapping in the hot wind. "My children, your journey has now begun. Behold the ark that will bear you to a new world."

The ship paused just short of the children and a gap appeared in the leading edge. The floating platforms moved into the gap and disappeared into the ship. Cobalt followed. The ship moved across to the platform and a ramp descended and touched down in front of Josh.

"Well, Josh Knight, you can be the first pilgrim to a new world." Mama Celeste stood beside the ramp.

Josh shook his head. "No." Big Foot picked him up and threw him over his shoulder. His screams echoed across the crater as he was carried into an ocean of evil.

Chapter 36

Ross pointed to his phone. "It's Cobalt." He touched the speaker button.

"I assume Major Robert Miller is in the command vehicle with you." Cobalt said over the phone.

"I'm here Anthony. Long time, no speak." Miller said.

"I'd be remiss if I did not tell you a story, Robert." Cobalt said.

"Not another one of your stories, Anthony. I can't afford to take a nap right now."

"No, Major, you cannot. I see your puny forces are arrayed outside my encampment. I will not waste time trying to convince you of my peaceful intents since you are set upon assaulting this compound. But, I will remind you that it will be you who must fire first. I have no desire to engage in a fire fight."

"I thought you were going to tell me a story." Miller growled.

"Ah, yes. The story. Thousands of feet below the serene surface of the ocean there is a world of cold darkness. It was once thought nothing lived in these depths. But there is one creature known as the black devil angler fish. It has a most mysterious attachment to its head. The attachment is a long stalk and at the end of the stalk is a small protuberance. Some say the protuberance resembles a worm or a small fish. In the cold depths of the ocean, the angler fish has developed the ability to make the very small tip of this protuberance glow. Imagine, Major, floating along in total darkness unaware of the mysteries of the deep. And then, ahead you see a tiny glimmer of light! It flickers. It moves. It beckons. And so, you swim toward it mesmerized by its strangeness; enticed by the promise of its mystery. And, as you arrive just inches from the glowing light, something behind the light moves and your last conscious thought is filled with the horrid image of huge, gigantic crystalline teeth and gigantic eyes filled with endless hunger."

"Nice story. What's your point?"

"My point, Major, is you were lured by the Roswell redux. It enticed you. It pulled you here with your puny army. But, that ship was but a prototype, a chip off the old block as you would say. Behold the Leviathan!"

The phone went dead and a shadow passed over them. Ross stepped out of the vehicle and held onto the open door. He swore and motioned to Miller.

"Do you see this thing?"

Miller tried to adjust his console screen and gave up and leaned down to look out the window. "Get back inside, Ross. Now!"

The Leviathan rose from the crater and moved slowly toward them. Its huge expanse blocked the sun. The underside of the disc had over a dozen bubbles of armored metal and thick glass from which gun turrets swiveled to point at Miller's army. Miller grabbed Ross and pulled him back into the passenger seat.

"Shut the door and hold on." Miller threw the vehicle into gear and spun around, flooring the accelerator. "Raptors, all raptors, retreat one klik then turn and engage!" He shouted into his earpiece.

The other vehicles spun and followed them, kicking up dust from the desert floor. A shadow sped over them and the Leviathan appeared right in front of them hovering a dozen meters off the ground. Miller threw on the brakes and they lurched to a halt.

"He's playing cat and mouse. Fine! All Raptors, engage!" Miller screamed and tapped the man behind him on the legs. Ross covered his ears as the gun above him roared as it threw round after round toward the Leviathan. Puffs of smoke erupted from the purple surface of the craft. When the smoke cleared, the surface was untouched. Each of the vehicles exhausted their munitions until only the TOW missiles were left. The Leviathan had experienced no evidence of damage. Miller frowned.

"On my mark, release all TOW missiles." He glanced at Ross. "I told you, you should have stayed in bed. Fire!"

All missiles rocked the vehicles as they sped toward the Leviathan. One of the turrets on the upper surface of the Leviathan glowed with green fire and a wave of lavender and green energy rippled through the air. The TOW missiles exploded in mid air and the wave continued. Miller threw the MAV in reverse and floored the accelerator. The wave caught the forward vehicles and they exploded in green and lavender smoke. By the time the wave reached Miller, it had weakened but it still carried a punch. The MAV lifted ten feet into the air and green lightning sizzled around the interior, frying the electronics. Ross' hair stood on end and the MAV crashed back to earth with a thud. His teeth rattled and he bit the inside of his mouth. Miller's head bounced against the dashboard and blood spurted into the air. The MAV was still in gear and when its wheels touched down, it lurched backwards over a rock and turn up onto the driver's side.

Ross spat blood from his mouth and climbed through the broken passenger side window. He fell to the ground and stumbled to his feet. The MAVs closer to the Leviathan were nothing but burning mounds of flesh and metal. The Leviathan lifted quietly into the air and with a loud sonic boom, disappeared into the sky.

Chapter 37

Cephas Lawrence heard the roar of the helicopter and wobbled through the door. He shuffled along in his pink bunny house shoes and shaded his eyes against the unusually bright sun light. The helicopter appeared over the top of the compound buildings and he waved. It banked and settled onto the ground near the edge of the crater. Theophilus Nosmo King hopped out of the helicopter and lumbered toward him. Cephas gasped in shock.

"Theo! You are alive!" Theo gathered him into a huge bear hug. Cephas' eyes burned with hot tears.

"You bet I am, Papaw. I ain't gonna let no Golem get the best of me. I made them stop and pick me up in Shreveport." Theo put him down. "Pink house shoes?"

"They once belonged to Magan Celeste, the leader of this wacky UFO church."

Renee joined them along with a man dressed in jeans and a black tee shirt. His short, black hair was shot through with gray streaks. His face was vaguely familiar. "You must be Josh's father."

Knight nodded. "Arthur Knight. No time for blame games. Where's Josh?"

Cephas frowned and glanced at Renee. "With the children. I'm afraid they are in grave danger." Cephas pointed out over the crater toward the far wall to the slash of blackness that marred the rock face. "Cobalt has built a new flying craft much larger than his other two. He took Josh and the children and is headed for his space station."

"Major Miller and FBI Agent Ross were supposed to stop them." Knight said.

"Ross?" Cephas raised an eyebrow and massaged his huge mustache. "God does indeed have a strange sense of humor. I heard explosions and gun fire on the other side of the north rim." He pointed at plumes of smoke drifting into the air. "If there was a battle, I fear your Major lost. Cobalt's ship is quite formidable. And where is Jonathan?"

"I spoke with Cassie. They made it to Axum and Jonathan disappeared with a monk guarding the chapel of the Ark of the Covenant. Cassie is furious but I'm angrier with her. She went after the ark!" Renee said.

"We can only hope that Jonathan finds the throne, yes?" Cephas said. "Hopefully he will make it in time to keep Cobalt from harming Josh."

"Then what are we gonna do?" Theo asked.

"Go after them." Cephas hobbled back toward the doorway from which he had emerged. "After Cobalt and the children left, I managed to escape from my tiny prison. I haven't explored everything, but according to a blueprint I found on one of their computers, there is a hangar under the compound and in that hangar there may be the remaining ship. Follow me."

He led them down a dark stairway and they emerged in a cavernous hangar. The far end opened out onto the desert east of the crater. One of Cobalt's ship sat quietly in the middle of the hangar. Just inside the doorway, a man sat at a desk and console. He jerked in surprise at the sight of them. He held up his arms only so far before they were stopped by the shackles that held him to the desk.

"Who are you?" Knight asked.

The man tried to mumble something. Cephas hobbled over. As he neared, he noticed the dried blood on the man's chin. A beep came from the console and the man's eyes filled with horror. He turned back to a keyboard and frantically tapped away. The beeping stopped and he sighed.

"Who is this?" Renee asked.

The man pointed to a monitor and began tapping in words. "I am Dr. Barnard, Cobalt's lead scientist."

"Why can't you talk?" Theo asked.

Barnard's eyes filled with terror and he pointed to his mouth with one of his shackled hands. He turned back to the keyboard. "Cobalt took my tongue."

Cephas stumbled back into Theo and shook his head. "I'm so sick of this! So tired of this evil!" He shouted.

Theo gripped his shoulders. "Calm down, Papaw. We gotta think clearly or we don't get Josh and Jonathan back."

Cephas fought down the anger and nodded. "What is that beeping, Dr. Barnard?"

The keys tapped. "Every five minutes I have to put in a code or the cooling system on the Sunstone production rises. In twelve hours, the Sunstones will have been completed and these shackles will open." He held up the shackles and Cephas noticed an electronic lock at each wrist.

"What happens if the cooling doesn't work?" Knight asked.

Cephas put a hand on his arm. "Cobalt put one of those sunstones in Josh's ear. He said if it reached a certain temperature, it would release all of its energy at once."

Knight tensed under Cephas' grip. "How many of these things are in the crater?"

Barnard tried to swallow and turned back to his console. "Enough to level most of the state of Nevada."

The beep startled them all and Barnard turned to another keyboard and typed away. The beeping continued and he groaned, hitting the delete button and trying again. The beeping continued.

"Calm down." Cephas said.

Barnard wiped sweat from his brow and put in the code a third time. This time, the beeping stopped. He typed on the other keyboard. "You have to stop him. He's gone to his Eagle's Nest with the new ship."

Cephas nodded. "We know. Do you know how to fly this ship?" He pointed to the smaller craft. Barnard shook his head and typed into the console.

"I do not and besides, I'm kind of busy."

Arthur Knight listened to the tapping of the keyboard and walked toward the craft. This ship was smaller than the one that had taken Vega. And yet, it still evoked in him an unmistakable sense of dread. He recalled the sickly sweet fragrance of the gas and the numbing music. The skin of the craft seemed to blur in and out of focus. For a second, he thought he saw something move within the skin!

"This thing is evil." Theo appeared beside him.

Knight jumped and his hand went protectively to his chest. Crazy! Why would Theo take the Bloodstone? Did he even know it was there? As his fingers strayed across the stone beneath the fabric of his shirt the stone pulsed with heat. Could it be? He reached out his right hand and touched the surface of the craft.

The face appeared beneath his hand and he jumped back. Theo stumbled into Cephas Lawrence who had joined them. The face bulged out of the ship's skin and studied them. Its lips moved.

"Help me! Free me!" It pleaded.

Cephas put out an arm and pushed Knight and Theo back. "I am Dr. Cephas Lawrence and in the name of my Lord Jesus Christ, I command you release this man." He shouted.

The face quivered and shook and skin bulged and stretched. The man fought and shoved and pulled until he fell away from the skin onto the floor. He was naked and covered with purple ooze. His hair was long and matted and a beard draped down to his chest. Cephas helped him to his feet. Renee hurried over with a fire blanket she had pulled from an emergency cabinet and wrapped it around the man. Theo helped him hobble to a nearby tool bin and the man sat down.

"Thank you! I've been in that insane thing for Lord knows how long. I'm Ralston Mead."

"You were in the skin of that ship?" Knight asked.

"Of course." Cephas looked at the ship. "It's not an inorganic metal construct. It's alive."

"Of course it is." Ralston coughed. "And, there are others inside. It absorbed us to try and understand who we are and where it was. I've lost all sense of time and space. Most of the others have gone mad."

"It's alive?" Renee said. "Then, the ship Cobalt made?"

"Must be alive, too." Knight nodded. "When I first worked with him, he was into genetic engineering. I've always wondered how he moved from that to space craft engineering. Now, it all makes sense. His new ship is sort of a clone of this one."

"And, the other one, too." Cephas said. He stiffened and glanced at Knight. "If this ship is alive, then it can be possessed."

"Evil." Ralston whispered. "When the pilot is present, it is evil beyond anything you can imagine."

Knight left them and walked toward the ship. If it was alive and if it could be possessed, then perhaps he could communicate with the ship through the Bloodstone. He reached into his tee shirt and pulled out the stone and let it dangle over his chest. He gripped it with his right hand and paused just inches from the skin of the ship. He reached out a hand and touched the blurry surface.

"Who are you?" He asked.

The skin grew mottled and changed from green to lavender. "I am in pain." A voice rasped in his head. Knight almost pulled away as wave after wave of pain crashed through his mind. "I am incomplete." The whisper continued.

"Who controls you?" He asked.

"No one, now. The one who hurt me is gone with the obscene one." The voice continued. An image of a dark, shadowy creature surfaced from the deepest, darkest recesses of Knight's mind. It quivered and lurched into the image of a huge ship purple and bruised and covered with pustules and bubbles of yellow energy. It was more than he could take and he stumbled back and fell to the ground. He blinked, trying to push the images out of his mind and Renee's face appeared in his line of sight. She leaned into him.

"Arthur, it's me. Come back to us." Her hand slid into his and he tightened his grip. He blinked away the confusion and sat up.

"It is alive and it has been under the control of a foul demon." He said hoarsely. "That demon now powers Cobalt's other ship." He glanced at Cephas. "It's huge!"

"I told you. It is an abomination, an amalgamation of this ship and our world and the world of evil he wishes to dominate." Cephas said.

They helped Knight to his feet. "I think I can fly this thing."

He hobbled over to the ship and with great reluctance, pressed his hand against the skin again. "Will you let us in? Will you let me pilot you? We have to stop that other ship."

"Yes! I need more power for propulsion, but I can let you control me since you wear the Heartstone."

Heartstone? Knight glanced down at the jewel. "We have a source of energy. The Sunstone. Will that do?"

"I sense these stones of which you speak. They are manufactured from the evil one, but they are not themselves evil. I can use them. Yes, I will help you stop the evil one but it will cost you greatly, Arthur Knight."

"How do you know my name?"

"Our minds have touched. Now that I am free of the evil one's control, my ability to see a little ways into the past and a little ways into the future have returned. If you do this, the cost for you will be grave." It whispered.

"I am willing to do anything to save my son and my daughter." Knight said.

"Such love is what I have craved. Such love is a power greater than anything the evil one has. I will release the beings the evil presence forced me to imprison."

A doorway opened in the dome on top of the craft and a ramp slid down to the floor. Knight stepped back as a group of people began walking down the ramp. There were over a dozen, all naked and covered with purple goo. Ralston Mead stood up and hurried over to them. "This way. Hurry!"

They followed him across the hangar floor and he motioned to Knight. "I can take care of my friends. Go! Stop Cobalt! For us!"

They left the ship's prisoners with Barnard and Cephas told them how to find food and clothing. It seemed the church had a new congregation! Once inside the ship, Knight found his way to the control room and the ship flew beneath the sphere over the crater. The cylinder filled with gleaming points of light. Sunstones! Cephas studied the consoles around them.

Cephas examined every surface. "It would seem that Cobalt has added some modern touches."

"The people he abducted were part of Area 51. They helped him construct additional modern technology to the ship. These consoles weren't a part of the original ship and there are four escape pods along the outer corridor."

"Four? That was generous." Cephas said.

Knight shook his head. "No. He wanted to make sure there was always one nearby. Now, I've told the ship to fly us over the north rim so we can see what happened with Miller and Ross."

A large view screen illuminated and they moved out of the hangar and out over the desert. Slowly, the ship flew along the outer rim of the crater. Renee gasped at the sight of what was left of Miller's decimated army. Two lone figures trudged across the desert from the smoking remains of their vehicles. Under Knight's instructions, the view magnified. Major Miller and FBI Agent Ross stopped and looked at them.

"Shall we pick them up?" Theo asked. "We ain't got a lot of time to take them back to an airbase."

Ross activated a communication interface and Ross' voice came over the console. "Go ahead and blast us, too!"

"Let me." Cephas leaned into the console. "I'm afraid we aren't going to be able to help you out, Agent Ross."

"Cephas Lawrence? Is that you?"

"Yes. We are headed after the ship that attacked you."

"Well, good luck with that! They cut through Miller's offense like a hot knife through butter." Ross said.

"Have you contacted anyone for help?"

"No. No phone signal."

"Renee has a helicopter at Cobalt's compound. We'll send it to pick you up."

Ross nodded. "You stop Cobalt and tell Steel he owes me big time."

"Who is there with you?" Miller asked.

"Arthur Knight, Theo King, and Renee Miller."

"Renee? Cobalt has Vega. And, Josh Knight. Be careful." He said.

"I will, Robert." She whispered. "I'll get our daughter back and keep her safe. I promise."

"Oh, and there is a Dr. Barnard who needs help in the compound with those sunstones." Cephas said. "Otherwise, this desert will see a nuclear blast unmatched since the 1950's hydrogen bomb tests.

Chapter 38

"Is that Cobalt's space station?" Steel asked.

"Yeah, the Eagle's Nest." Vivian slowly stood up and tried to straighten her hair. "The sight of it makes you all tingly inside, doesn't it?"

Steel ignored her, mesmerized by the sight of the space station hanging against the black velvet of space. An encircling ring of golden metal and glass composed the main body of the structure. It must have been ten stories tall and a city block in width from its outer rim to the inner rim. Its upper edge was larger than its lower edge so that a cross section would resemble in inverted trapezoid. Sitting above the space station a flattened sphere of metal and glass was held in place by gleaming golden struts and resembled a huge magnifying glass. Beneath this object a second ring was suspended made of silver and glass. In the exact center of the second ring was a large, flattened crystalline sphere and Steel could see people spinning and wheeling their way in the zero gravity. The entire structure was three football field lengths in width and it hung against the incredible blue of earth like an elaborate earring. Higher above the station in the far distance a tear shaped, copper colored structure glittered in the sunlight.

"That's the Solar Mirror Cobalt has been working on for years." Vivian said. "When the mirror opens, Cobalt plans on capturing a huge amount of solar energy and feeding it directly to the Eagle's Nest."

Steel tried to ignore her and focus on the space station. Somewhere inside was Josh Knight and possibly Cephas Lawrence. How was he going to escape a space station? The ship flew beneath the near edge of the space station and Steel gasped at the waves of evil that met them. He stepped back away from the window and landed on the couch. Vivian leaned against the glass and giggled at the sight of what lay before them.

"He did it!" She whispered.

"What is that thing?" Steel asked.

Another ship a third of the diameter of the space station hovered beneath the station. Its knobby, irregular surface was covered with metal appendages and bruise colored knobs. Steel grimaced and squinted as if looking into bright sunlight. "It's obscene."

"Yes." Vivian sat beside him. "I don't care if you know this, Steel. I'm scared out of my wits. That thing is horrific! I had no idea this is what he was building. No wonder the Council wants him to fail."

"Then we have to stop him." Steel said hoarsely as their ship moved past the huge abomination. Between the larger ship and the flattened sphere, a third disc shaped chamber was suspended by long, translucent tubes to the lowest levels of the station. Their ship flew into a rectangular port on the edge of the lower ring and came to rest inside a hangar.

"Then follow my lead." Vivian said. "Rey, baby!" The nightfly appeared outside the open door to the lounge. "Take Mr. Steel to Anthony Cobalt. If you have to, use the inertia gun on him."

Steel grabbed the canvas bag from its hiding place and shrugged into it. The nightfly didn't seem to care and it escorted Steel off the ship and across the hangar to an inner chamber. In spite of the absence of any centrifugal force, there seemed to be normal gravity. Steel glanced around him as other nightflies moved in jerky motion around the hangar, moving pallets of supplies down a long metal tunnel that led toward the distant ship.

"The Leviathan." Rey said.

"What?"

"It is called the Leviathan and it will take Cobalt's army to the Promised World." Rey motioned with the inertia gun toward a doorway leading toward the center of the space station.

"Will that include you?"

Rey activated a switch and the doorway slid open. A long, transparent tube led away from the outer ring of the space station toward the circular, flattened chamber hanging between the Leviathan and the sphere of floating people. Rey pointed to the tube.

"There is no gravity. Pull yourself along the railing to the central chamber."

Steel stepped through the threshold and was instantly floating in space. The sensation was confusing and he fought dizziness. He managed to snare a nearby railing and anchor himself with his head toward the distant chamber. His grip tightened as he fought the sensation of falling. Rey sped past him on his four wings. He pulled himself along the railing and control over his body returned. He caught up with Rey.

"You didn't answer my question."

Rey paused and his vibrating wings held him motionless. "I am defective. I was the first of Cobalt's genetic modifications. My demonic spirit will only attach to this body so I had no hope of a new one. Until Cobalt's goal is reached, I am a slave to this body." He spun and continued forward.

Steel reached the doorway leading off the tube. The disc like chamber was made of copper colored material and the inner doorway slid open. Gravity returned inside the door and Steel lurched under its sudden control. The chamber beyond was circular with a metal floor around the periphery and metal walls. But, the center of the chamber was transparent allowing an excellent view of the Leviathan beneath. Nightflies crawled over the ship's surface like flies on dung.

The ceiling was transparent allowing him to see the sphere suspended directly above the chamber's center. Inside the sphere, he could see people wheeling and spinning in zero gravity. But, the object that commanded his attention sat at the center of the floor. The black, rocky skin of the Fallen Throne seemed to soak up light as it sat at the edge of a gleaming metal disc. The Infinity Disc glittered in reflected sunlight.

"It is indeed a breathtaking sight, don't you agree Mr. Steel?" He whirled. Anthony Cobalt floated out of the transport tube. He wore a gleaming white one-piece coverall with silver highlights and a silver cape. In his hands he bore a huge, crimson jewel the size of a human head. He landed deftly on the floor and strode toward the throne.

"If you'll follow me, Mr. Steel, I will let you witness an event unparalleled in human imagination."

The cold snout of the inertia gun touched his arm. "Follow." Rey said.

Steel fought panic as he walked across the sheer transparent floor. The Earth gleamed beneath him and the Leviathan glowed with bits of lavender and green points of light. "I suppose the gravity is supernatural?"

Cobalt paused at the edge of the Infinity Disc. "Of course. But, the space suits are real." He gestured to a rack of white and silver space suits nearby. "The thirteen empty slots once held suits for each of the children. And a few others for safety's sake."

"Why don't you take a walk in the vacuum?" Steel paused beside him.

"Humor does not become you, Mr. Steel. Now, if you would be so kind as to put on a space suit."

"Why?" A nightfly brought a spacesuit over to Steel.

"You might have to talk a walk in the vacuum of space." Cobalt said.

"And, why should that matter to you?"

"I might need you alive, Mr. Steel. Leverage. Only a fool kills his enemies before the events are finalized. Now, I can have a nightfly zap you with the inertia gun and then stuff you into the space suit, not carefully I can assure you. Or, you can do it yourself and be certain all appendages are in their proper place."

Steel jerked the space suit from the nightfly and stepped into. It took some doing, but he managed to get his arms into the sleeves and sealed the suit. The nightfly placed his helmet on the floor next to his feet.

"Very good, Mr. Steel." Cobalt glanced at his chronometer.

"Soon, the children will arrive. And, the Event is only a few minutes away from reaching earth."

"The Event?"

"A CME." Cobalt smiled. He walked around the Infinity Disc and deposited the Bloodstone into the cradle behind the mask. It barely fit as most of the throne was coated with black stone. The Bloodstone began to pulse and glow with energy.

"A coronal mass ejection from the sun. It will be instrumental in completing my plan." Cobalt stepped back. "But, for now, the Bloodstone contains enough stored energy to free the throne from its lava coating."

The black stone of the throne began to crack. Red light shone from within the cracks and with a sudden explosion, shards of black rock shot away from the throne. Cobalt held up his hand and the pieces of stone floated harmlessly away toward the far wall. The throne gleamed with azure light.

The Bloodstone settled behind the mask and the eyes began to glow. The mask and stone changed shape and became a seamless crimson head with slit eyes and fiery hair. The head turned and regarded Cobalt.

"Long have I waited. Long have I been imprisoned." A haunting voice echoed in the room. "I am freed by the Bloodstone."

The throne began to glow with blue light. The light flowed along the edges of the wings and into the arms and legs. When the light touched the disc, like water it flowed along the etchings in the metal. Steel tried to read the etchings but they blurred in and out of focus.

Cobalt sighed deeply. "Finally. The Fallen Throne is restored and reconnected to the Infinity Disc. All I lack is the sword. Which you have conveniently supplied, Mr. Steel. Reyjacklic, please take the canvas bag from Mr. Steel."

Steel had placed the bag on the floor while putting on his suit. The nightfly grabbed the strap of the bag and tried to hand the bag to Cobalt and the man stepped back. A sudden wave of fear crossed his features.

"What's the matter, Cobalt?" Steel asked.

"It is of the Light and I cannot touch it. Where is Vivian?" He asked Rey.

"She is coming."

"When she arrives, instruct her to place the sword in the hands of the throne."

Steel glanced at the tunnel leading back to the hangar. In the distance, he saw Vivian pulling herself along the railing. "That's why you took her demons, isn't it? So she could handle the sword."

Cobalt lifted an eyebrow and ignored his remark. "Are you familiar with a story from the Bible, Mr. Steel?"

"There are lots of stories in the Bible."

"Ah, but this one is a parable. A tale of a man who was inhabited by an evil spirit. The spirit left the man and wandered through waterless places seeking rest. Always seeking rest where there is none; nor will there be any rest forever, Mr. Steel. It finds no rest so it reasons to itself that it will return to the house from which it came. And, when it returns to that house, it finds it empty but swept and clean and in order. Gone is the chaos and confusion. It's a perfect fixer upper ready for a new tenant. So, what does the evil spirit do? Why, it goes and finds seven other spirits more evil than itself and they all enter this house and they dwell there and the last state of the person is worse than the first!"

"What if she doesn't want the seven demons?"

"She has no choice." Cobalt stepped closer and smiled. "For you see, Mr. Steel, if the house is in order and swept and cleaned and orderly but it is still empty, then there is no barrier to the returning spirits. Nature abhors a vacuum, even on the spiritual level. Vivian craves the return of her demons. She is empty and has nothing else to fill herself with."

"There is always, as you put it, the Light."

Cobalt laughed. Then, he laughed some more. "How naïve and innocent you are to believe that everyone wants to commune with God. Have you ever stopped to think about the fact that maybe, just maybe, there are those of your kind who hate God so much, they would rather spend eternity in Hell than spend a second in the presence of God? You see,

Hell is not a punishment for man, Mr. Steel. It is a choice. *The* Choice. Ah, here's Vivian now."

Vivian stumbled as she stepped across the threshold into the gravity field. She walked across the vast chamber and her step was slow and weak. Her eyes were rimmed in dark circles. She wore one of the white jumpsuits and she paused beside Rey and leaned forward with her hands on her knees.

"I'm getting weak, Cobalt." She gasped. "I've done what you asked. I want my demons back."

"Not yet, dear Vivian. There is one last task. But, before you complete your task, let's have the children." He motioned to a doorway opposite the one they had entered from and the children appeared from the tunnel. They wore white space suits and even from this distance, Steel thought they looked huge. Vega led the group and when she landed in the gravitation field her eyes widened.

"Hot Steel? Is that you?" She ran around the Infinity Disc and grabbed Steel up in a huge bear hug and lifted him off the ground. "I thought I'd never see you again. You look older."

"And, you've grown." Steel grunted as she put him down.

"Now, Vega," a stately woman in a flowing gown appeared beside Cobalt. Her eyes glowed with power and her bare head matched Cobalt's. "You promised you would cooperate. Please join the others at their stations."

Vega frowned. "Magan Celeste, this is my friend from long ago. He saved me from the evil Captain."

"Well, now you have shown him great respect and gratitude. Your station. Now." Magan Celeste fought to keep her smile.

"I'm growing tired of listening to you, Magan Celeste. You promised me Mrs. Donnelly would be back."

"I told you to be a leader. Follow our orders or some of your friends will be hurt." Magan Celeste frowned.

The heat came off her as Vega moved toward Magan Celeste and Cobalt. She towered a foot taller over them and her eyes filled with power. "You will NOT hurt my friends. One of them is your own daughter. You know we have powers. We could use them."

"And, then you would be no longer good." Cobalt said quietly. "You want to be a good person, don't you Vega? Well, if you use your powers to hurt someone like Bobby did on the mountainside, you would become a bad person."

Vega froze and looked at Steel. "Bobby didn't know what he was doing. He had no idea it would hurt those men."

"I know, Vega. The Captain manipulated all of you. But, he's not here and I think it is best to cooperate for now." Steel said.

"I do have one other incentive, Vega." Cobalt said. He pointed to another tunnel. "We have some more friends who have just arrived."

Steel whirled and from the nearest tunnel escorted by a host of nightflies Cephas Lawrence and Renee Miller stepped into the gravity well. They each wore one of the white space suits.

"Mother!" Vega ran across the chamber and scooped Renee up in her arms. The nightflies scattered away from Vega as the two of them embraced. Cephas hurried to Steel and grabbed him around the chest.

"Thank God you're alive. I expected the worst." He said.

Steel awkwardly embraced the old man. "What happened to you?"

Cephas pushed back and swiped at his eye. "Vivian and Cobalt came to the house. They have Josh somewhere."

Steel glared at Cobalt. Cobalt smiled. "Yes, where is Josh? The answer to that question is the only reason your friend, Theo, hasn't pulled me limb from limb, Dr. Lawrence."

"What have you done with Theo and Arthur?" Cephas asked.

"They are still on the ship, Dr. Lawrence. I had no idea Arthur Knight would be able to pilot the ship with his small fragment of Bloodstone. It is best he keep his distance for now and Theo is far too violent to be trusted not to interrupt our coming ceremony. Now, that our family unit is restored," he gestured to Renee and Vega, "We can get on with the festivities. Vega, will you please join the other children at their stations? Before you refuse, please realize I have ample leverage to guarantee your cooperation, do I not?"

Renee and Vega joined Steel and Cephas. Renee patted Vega's arm and tried to smile. "It's okay, Vega. I'm here now. Do as the man says."

Vega nodded and walked around the Fallen Throne. The other children were lined up along the far edge of the Infinity Disc and she joined them. The nightflies brought a helmet to each child.

"You will hold onto your helmet for now, children." Cobalt said. He gestured to the transparent ceiling. "If you will turn your eyes to the starry realm, you will witness the blossoming of my beautiful sun flower. Watch it closely, for as the flower blossoms, the sun will fall upon us all."

"The sun?" Cephas asked as he squinted and tried to focus on the distant copper colored object far above the Eagle's Nest. "Your solar mirror, isn't it?"

"Yes, Dr. Lawrence. When the mirror unfurls, it will collect energy from the sun and focus it upon this space station. In fact, the Event approaches. An eruption from the sun. When a coronal mass ejection hits the Earth's magnetic field, it is compressed on one side of the Earth and creates a tail on the other side of the Earth. A CME can create trillions of watts of power. Can you even imagine that kind of power, Dr. Lawrence?"

Cephas wiped his mustache and stepped toward Cobalt. "But, that much energy passing through this space station will kill us all!"

"Not if it is focused into a tight beam." Cobalt pointed to the Fallen Throne. "That beam of energy will fall upon the Bloodstone with an influx of unimaginable power." Cobalt gestured to the throne. "Vivian, if you will take the crystalline sword from its bag and place the hilt in the hands of the Fallen Throne angel."

Vivian slumped forward. "Then, will you give back my demons?"

"Of course, dear. Now, hurry along. The Event is moments away."

Vivian pulled the sword from the canvas bag. It was huge and dull. She dragged it across the floor and stepped up onto the Infinity Disc. She glared at Cobalt one last time with such hatred and revulsion Steel was surprised she didn't use the sword to decapitate the man. She lifted the sword above her head and almost stumbled. Stepping up onto the seat of the throne, she seated the hilt in the hands of the angel statue, her face just inches away from the glowing eyes of the thing's face.

The sword glowed with blue light and the angel spoke. "I am now complete. What is your bidding?"

Vivian slid off the throne. "My bidding?"

"No, the bearer of the Bloodstone."

Vivian stumbled from the Infinity Disc and back to Cobalt. "Now, I want my demons back."

Cobalt laughed and gestured to Reyjacklic. "Reyjacklic, take Vivian to the Worship Sphere, please."

"What?" She reached for him and two nightflies grabbed her arms and pulled her away. "You promised!"

Cobalt shrugged. "I lied. Reyjacklic, obey me and use the inertia gun on her. Take her to the Worship Sphere with the rest of the congregation."

Vivian jerked her arm out of the grasp of the nightflies and stumbled into Renee Miller. Renee grabbed her and they spun away from Cobalt.

"Get your hands off of me!" Renee spat.

Vivian leaned into her and mumbled something. Then, the inertia gun was glowing red and Vivian froze. Reyjacklic picked up her paralyzed body and moved toward the tunnel.

"What is the Worship Sphere?" Cephas asked.

Cobalt pointed to the sphere between the chamber and the far solar mirror. "It is the focusing point, Dr. Lawrence. All of the energies from the sun will pass through the Worship Sphere and will then be focused on the Bloodstone."

"But, there are people in the sphere." Steel said.

"Yes, Mr. Steel. I believe they will become what your culture might call crispy critters."

Chapter 39

Josh Knight should have been having the time of his life. He was in space! Not only that, he was floating weightless in a huge transparent sphere. He should have been fascinated with the Earth floating beneath him and the massive space station around him. But, his exuberance was tempered by the odor of vomit mixed with incense. Most of the worshippers were enjoying the zero gravity. About thirty were huddled against a railing along the edge, crying and vomiting into the air. Amoebas of vomit floated around them. He took some pleasure in noting that one of the sick people was Big Foot.

He still reeled from the evil of the Leviathan. Big Foot had held him down until one of the nightflies had shot him with some kind of gun that paralyzed him. Big Foot had placed him in front of a viewport so he had to witness the destruction of Miller's small army and then the arrival at the Eagle's Nest. Big Foot had taken his paralyzed body and thrown him into what was called the Worship Sphere and it was only then his strength returned.

He "swam" his way through the air to the portal connecting to a tunnel. He grabbed the railing and waited. Eventually, someone would have to come through the doorway and when they did, he would make his way back to the hangar area and find Jonathan. Beneath him, the Leviathan crawled with nightflies and even from this distance, he felt the evil. Between him and the ship was an odd, disc shaped chamber of copper metal and transparent ceiling. In its center he could make out a throne shaped structure. Was this the Fallen

Throne? If so, then Jonathan was here for sure. But, where was Uncle Cephas? And, more importantly where was his father?

Someone floated up to him and he glanced into the face of Mrs. Donnelly. She smiled. "It's a beautiful site, isn't it?"

"You're here! Get me out. Now!" He whispered. "And, no that ship is not a beautiful sight."

"I was referring to your home. God created it just for humanity. A wondrous jewel hanging in the darkness of space. Do you realize how rare it is? And, Cobalt wants to leave it behind for a promise of a new world he thinks is better."

"Why doesn't he just fly off to this new world?"

"Because his kind is trapped, Josh. The Barrier keeps the fallen on Earth. They can still move in and out of the heavenly dimensions but physically, they are confined to this world. It is their curse."

"And, ours, too."

She turned jade green eyes on him. "But, it was your choice. Man once walked with God in paradise, but he chose the knowledge of this world; the knowledge of this universe; the knowledge of good and evil and life and death. He chose mortality and this space-time continuum over the heavenly realm."

"Why do I have to suffer for one man's choice?"

"Because you would have made the same choice, Josh. You have to know. You crave to be God. You are made in the image of God, the 'imago dei' but you chose to become more like God and not rest in your comfort. This is the results and you must live with the presence of good and evil every day. All about you." She reached out and touched his arm. A warm energy flowed through him bringing him comfort and strength.

"You must trust God, Josh. There are painful moments ahead but He is still in control."

"I want to see my father."

"You will. Be patient." She swirled in the air and disappeared from view.

Vivian slumped against the wall after pulling herself along the tunnel. Rey hovered before her brandishing the inertia gun.

"I know what happens to the people in the Worship Sphere, Reyjacklic. Cobalt has betrayed me." She said.

"He has betrayed us both. I am not to make the Journey to the new world with the others. I am defective." He lowered the gun. "What shall we do to stop him?"

Vivian's heart raced and she glanced up at the nightfly. "You did pledge yourself to me?"

"Yes."

"I shall honor you. You can become one of my demons and when the day arrives that I rule the Council of Darkness, you will have a seat." She found new energy and stood up. "I passed off the Bloodstone ring to Vega's mother and told her to use it to stop Cobalt. But, if we go back, there are more nightflies to stop us."

She paced along the corridor. "Why does he want to kill the people in the Worship Sphere? We both know that human sacrifice is a useless tool and is only to cement one's devotion to Lucifer. He knows that." She stopped and recalled a memory the ring had shown her. In that remote memory was Kabal's first encounter with the Fallen Throne when it opened a Portal to the Promised World. What had the Fallen Throne angel said? Something about a sacrifice would be required to reopen the Portal? She swore. She glanced out a nearby viewport at the distant mirror satellite, it's petals unfolding like some obscene flower. Then she glanced at the Leviathan. Cobalt needed the people in the Worship Sphere to die. He needed them in some way to open the portal large enough and long enough to get the Leviathan through. She whirled.

"The two ships that are left? Where are they?"

Rey blinked his cow like eyes. "Both are in the hangar."

"But, I need a demon to power the ship." She bit her fingernail. "And, I am afraid you are not powerful enough, Rey."

"One of the ships arrived with four humans. One of those humans flew the ship using his own power."

"What? How?"

"He has a shard of the Bloodstone."

"That's right. Cobalt said so. Wait, Arthur Knight? Josh's father? And, Josh is in the sphere. Rey, I hate to admit it but I'm about to rescue two hundred people from certain death."

Rey stepped back. "This is not like you."

"Not at all. But, it will stop Cobalt and that's all I care about. Come on." She headed for the hangar.

Chapter 40

Arthur Knight paced back and forth around the paralyzed figure of Theo. He had tried to attack the two nightflies guarding them and he had been instantly immobilized by an inertia gun. "I tried to tell you. But, no, you wouldn't listen, would you?" He shouted at Theo. "You just had to go barreling in without a plan and now, they've taken Cephas and Renee because you wouldn't behave."

He hurried back and forth and drew closer to the nearest nightfly. If he could catch the thing off guard and grab his inertia gun before the other nightfly reacted, maybe they could get free.

The door to the hangar slid open and another nightfly appeared carrying an inertia gun. He shot both of the other nightflies in rapid succession and they tumbled like overturned statues. Behind the nightfly, Vivian Darbonne sauntered into the hangar.

"Boys, I'm here to rescue you." She said. "What's up with the big guy?"

Arthur rushed her and at the last minute, the nightfly stepped between them. "Don't make me zap you." It said.

Vivian smiled. "Do you want to see your son alive again? Then, don't fight me. I'm on your side for now and we have to rescue him before Cobalt burns them all."

She took the inertia gun from the nightfly and twisted a dial. She aimed it at Theo and he relaxed. He took one look at her and balled up his fists. "You gonna taze me again with that intergalactic zap gun?"

"Honey, I live to taze you. But right now, we need to

cooperate. If Knight here can fly this ship, we have about two hundred people to rescue, including Josh. Now, are we going to fight about this or are we going to get busy?"

Theo glanced at Knight and he nodded. "I don't trust you but if you say you can save Josh, I'll not fight you. Problem is, this ship won't hold that many people. Maybe twenty or thirty at the most."

Vivian swore and planted her hands on her hips. "We need both of them, don't we?"

"I can fly this one." Knight pointed to the ship behind him. "But, who will fly that one?"

Vivian slumped and seemed to deflate. "I know someone I can ask to help, but he will won't something in return."

"Then give it to him." Knight said. "You want our help, get someone to fly that ship."

"We don't have much time. Maybe if you can get as many off the ship as possible, it might be enough to stop Cobalt's plan."

"Where are we going?"

Vivian studied the empty cylinder in the control room of the larger of the remaining two ships. "Rey, you don't have enough power to fly this ship, right?"

"No, mistress." Rey said.

She sat at a console and closed her eyes. "I never thought I would say this, but Bile I need you. Now!"

The air behind her popped and she turned. Bile stood before her in a three-piece tuxedo. His hair was wispy and thin. The spiral pulsed around his right eye. "Do you realize how annoyed I am right now? I was in the middle of something very important and you summon me because you have messed up. Royally!"

"I need you to fly this ship. If I am to stop Cobalt, I have to get two hundred people out of the Worship Sphere before he kills them all."

Bile studied her warily. "Where are your demons?"

Vivian stood up shakily. "He took them."

"You're empty, aren't you? Vulnerable?"

"Yes." She whispered.

"Why should I help you?" Bile crossed his arms.

"They are worshippers. They followed Magan Celeste and her church teachings. You could tell them you are a spirit guide sent by Celeste. You would have two hundred faithful followers at your beck and call." Vivian tried to smile.

Bile nodded. "Not bad, Vivian. You are an expert temptress. Very well. Let's get this show on the road."

"I know what the Fallen Throne is capable of when it is joined to the Disc and the Bloodstone." Steel said.

Cobalt turned away from the sight of the distant solar mirror. "Really? And, how would you know such?"

"Your enemy showed me. Donnatto, right?"

Cobalt scowled and looked away. "I wondered when he would show up. He always tries to spoil my fun. But, this time, there is no stopping me."

Cobalt motioned to his nightflies and they prodded Steel, Renee, and Cephas closer to the edge of the Disc. "The Coronal Mass Ejection is already on its way here. In nine minutes It will strike the solar mirror and power will be directed through the Worship Sphere above and into the Bloodstone. The timing is critical. So, we must get started."

He reached out and took Magan Celeste by the arm. "Magan Celeste, what have I done for you?"

Magan Celeste looked at her feet. "You have blessed my church."

"I mean you personally."

She looked back at him and he waved his hand over her head. She gasped as she seemed to shorten. The brown spots on her head changed into knobs of flesh. She bent suddenly in pain and her face was deformed with dozens of protuberances.

"Café au lait spots." Cobalt pointed to the brown spots on her head. "Magan Celeste suffers from neurofibromatosis.

An affliction her daughter does not share thanks to the intervention of the Bloodstone."

"Master, please!" Magan Celeste managed through thick lips. "It hurts!"

He waved his hand over her again and in seconds, she returned to her previous appearance. She sighed deeply. "Thank You."

"I bring this up, my dear to remind you. You owe me your very life. And, in gratitude to me for this healing, you have been a faithful follower for years building my church and gathering my congregation; my sheep. It is only fitting that you should be the first to sit on the Fallen Throne."

Magan Celeste blushed and tears began to trickle down her cheeks. "My lord, you are too good."

"First, we must have the children put on their helmets and activate their space suits." He motioned to the attending nightflies. Vega protested at first but then relaxed as her helmet was placed over her head and sealed.

"Why space suits?" Steel asked.

"In case there is a leak, Mr. Steel. We are in the vacuum of space. Keep your helmets close by." Cobalt extended an arm to Magan Celeste. "Would you accompany me to the throne?"

She linked her arm to his and he led her across the Infinity Disc to the foot of the Fallen Throne. He positioned her to face the head of the angel. Quickly, he stepped away and off of the Disc. "Now wait for the angel to address you." Cobalt said.

The angel's blue, crystalline face pulsing with the red eyes of the Bloodstone tilted in her direction. "Why have you come?"

"I wish to sit on the throne." Magan Celeste said as she wiped tears from her eyes.

"Do you wish to be one with all of time and space?" The angel's voice boomed through the chamber.

"Yes."

"Do you want to become one with the Veil?"

Magan Celeste cast a questioning look at Cobalt. He smiled. She nodded. "Yes."

"Do you give your body and soul to the universe?" The angel asked.

"Yes!" Magan Celeste said in sudden passion. "Yes! Yes! Yes!"

"Take your place upon the throne." The angel said.

Cobalt glanced at his chronometer and held up a hand. "Just a moment, Magan Celeste. We must time this perfectly." He waited and then lowered his hand. "You may claim your reward."

Magan Celeste turned deftly and had to stand on her tiptoes to reach the seat. She sat slowly, her eyes closed in ecstasy and then grabbing the throne's armrests, slid back until her head rested just beneath the angel's head. She pressed her head against the angel's breastplate and sighed. "I can feel the energy flowing through me. I am one with," She paused and horror filled her face. "What? No!"

The angel's lower set of arms pulled free of the throne's armrests and encircled her waist. Magan Celeste looked down and pulled against the blue crystal with all her might. "No! Don't let this happen!"

The lower wings moved and wrapped themselves tightly about her knees and lower legs. Then, the upper wings moved with sudden swiftness and wrapped around her upper torso. Steel reached out and grabbed Cephas and Renee afraid they might try to help. Through the translucent crystalline wings, they watched her struggle with the constricting arms. Then, the upper set of arms bearing the crystalline sword lifted higher, turning the blade downward, pointing its tip at Magan Celeste.

"Your sacrifice is accepted." The angel bellowed and the sword thrust down with sudden swiftness. The blade entered Magan Celeste's chest just below her chin and her scream was cut off as it severed her windpipe.

"Mother!" Even through the space helmet, one of the children screamed. A cylinder of lavender light erupted from the base of each of the children's stands and they were instantly paralyzed. Blood ran from beneath the wings of the Fallen Throne angel and began to trickle into the indentations on the

Infinity Disc. The blood glowed red and the Bloodstone pulsed with growing power. The blood flowed quickly circling the Disc and ending up surrounding the central indentation. Blue light erupted from the center of the disc and with a snap that popped Steel's ears, the mirrored surface of a perfect sphere formed. It grew until it filled over half of the diameter of the disc. The mirrored surface cleared and the sight before them drew gasps from Steel and Cephas.

Through the hemispherical surface of the disc a planet was visible dotted with green islands and continents separated by crystal blue water. Clouds swirled over the land. And, in the distance, lights gleamed from the dark side of the planet.

"Behold my Promised World!" Cobalt said.

"Why did she have to die?" Steel asked.

"You should know." Cobalt stepped toward him. "I do not understand how, but you had a vision, did you not? You saw the day I lost my access to the Promised World."

"I did."

"There is a gap between here and there." Cobalt paced around the slowly expanding sphere. "Remember the story of the beggar and the rich young ruler. They both died and the rich young ruler looked across at the beggar in the bosom of Abraham and asked for but a drop of water. And, the beggar said the chasm between his realm and that in which the ruler found himself was too wide. It could not be crossed." Cobalt gestured to the Fallen Throne. "But, there is a way to bridge the dimensional gap between this universe and the heavenly realm. And, it is through the heavenly realm one can journey to another point in space."

"A wormhole?" Cephas said.

"No, Dr. Lawrence. A wormhole is natural. We are talking about a purely supernatural phenomenon. And, there is only one way to pierce the Veil that separates this reality from the other. When a soul leaves the body, it breaks the dimensional barrier and leaves, for but a microsecond, a tear in the fabric of the inter-dimensional barrier. The Bloodstone's power was designed to keep that hole open long enough to create a stable portal between two points in space."

Cobalt motioned to the nightfly nearest to the last paralyzed child. "It is time."

The nightfly activated a switch and the thirteen pedestals were joined by a metal cable. Three nightflies tugged on the nearest child to the sphere. Steel tried to stop them, but a nightfly gestured with his inertia gun. The first child and his cylinder of lavender light intersected the sphere. With a sudden flash of light, the child popped through the surface and like a roller coaster on the downhill run, the other children followed. Renee tried to run to their side, but the nightfly stopped her with a burst of the inertia gun. All thirteen children now floated in the vacuum of space above a world somewhere on the far side of the galaxy.

"Dr. Miller, you should rejoice." Cobalt said. "You know your daughter cannot possibly survive on this planet. She is a Nephilim and will one day reach the height of 20 feet. Her only chance, the children's only chance is on another world. I can assure you, a suitable world exists for them or they would never have been allowed to exist by the creator."

"You can't just let them float in space!" Cephas said.

"It is not my intention." The chronometer flashed on his wrist. Cobalt raised his wrist to his face and the blinking chronometer painted his face in red light. "And, now for the Event! Behold, the power of the Sun." He smiled at Steel. "Beats the power of your Son any day!"

Chapter 41

Arthur Knight studied the view screen in the ship's control center. Vivian had communicated with someone who could power the other ship. He and Theo had flown this ship and stopped beside the outer surface of the Worship Sphere. "How do we get those people out of the sphere?" Knight said.

"I do not have the appropriate docking mechanism to interface with this inorganic device." The ship's voice echoed in his mind.

Knight saw people clinging to the railing around the perimeter of the transparent chamber. "Go to the nearest tunnel attachment. Maybe we can cut a hole or something."

"And, the air will squirt out into space along with the people, sport." Theo said.

"Then what can I do?" Knight slapped the console in frustration.

"Warning. The solar satellite is being activated by an incoming coronal mass ejection." The ship's voice said out loud. "The adversary Cobalt called this the Event. In three minutes, a highly energetic beam will traverse this chamber and all organic life forms will be reduced to ash."

"Okay, okay. Are there any primitive devices attached to you that I can use to fly over to that tunnel?" Knight asked.

"There are four escape pods on this craft. They are maneuverable."

"Theo, stay here. I'm going to rescue Josh, if nobody else." Knight ran from the control room. He paused before a white metal door in the corridor and pressed a large, green pad. The door swung open and he stepped into the escape pod. Six seats with safety harness occupied the room and one seat sat before a control console and view screen. It activated and Knight hurried into the seat, fastening his safety harness.

"I can activate this device for you, but you must pilot it once it has detached from my direct connection." The ship said.

Knight looked at the controls before him. A keyboard and a joystick. "Detach. Now!"

There was a thud and the escape pod ejected away from the craft. In the view screen, the pod pulled away from the ship. Knight pressed the joystick to the right and the escape pod rotated and brought the Worship Sphere into view. But, his forward momentum was too great and the escape pod impacted the wall of the translucent tunnel. A fissure appeared and the escape pod tore into the tunnel and came to rest against the far side.

Air screamed around the outside of the pod and he heard a thud on the outside. He opened the escape pod door and watched in horror as some of the worshippers were sucked through the opening into space.

"Dad!" He heard a scream. Josh Knight clung to the edge of the escape pod. He grabbed his son's hand as the air began to grow thinner. He pulled with all of his strength and Josh fell in to him. He jerked him back into the escape pod and slammed the hatch shut.

Josh gasped for breath and Knight hopped into the control chair and pressed the joystick forward. It tore through the far tunnel wall and out into black space. He piloted the escape pod with jerky motions underneath the collapsing tunnel. Suddenly, Vivian's ship shot into view and an opening in the dome on its roof engulfed the severed tunnel. A steady stream of worshippers poured down the collapsing tunnel into the ship.

A sudden flash of light and a gush of flames shot down the tunnel toward Vivian's ship and the dome slammed shut, pulling away as a dozen or so people shot out of the tunnel into space. The vacuum snuffed out their flaming bodies in a microsecond. The escape pod rotated and the Worship Sphere came into view. Light as bright as the sun poured into the top of the flattened sphere and still there were dozens trapped inside. They became human torches burning in the endless night of space. The light focused and emerged from the underside of the sphere as one coherent beam of energy that shot into the flattened disc chamber.

On the console before him a button blinked back and forth with two alternating messages. "Return to Earth" or "Return to ship". He pressed the "return to ship" button. It was only then he felt a hand on his shoulder and he looked up into the tear stained face of his son.

The entire space station shook with the impact of the energy beam from the solar satellite. Steel watched in horror as it flooded the Worship Sphere with yellow orange light and blobs of flame appeared among the floating people. "You're incinerating them!" He screamed.

"Yes, and each soul released from its body will widen the tear in the Veil." Cobalt's eyes widened with insane glee. "And, behold the power of the solar satellite."

A beam of red energy shot through the ceiling gap and impacted the Bloodstone. It shook with power and glowed hot. The sphere began to enlarge, spreading toward them across the floor of the chamber. As the interface between realms touched the metal floor, bits and pieces broke away and floated into the space of the foreign world.

Steel grabbed Cephas and pushed him back. Renee, freed of the inertia gun fell back with them as the sphere enlarged. The hiss of escaping air underscored the roar of the Bloodstone as its power output increased exponentially. Steel motioned to his helmet. "Get them on! The air is escaping."

"By all means!" Cobalt floated upward into the air above the sphere. A shimmering field of energy surrounded him. "I don't need a suit."

Steel sealed his helmet and made sure Rene and Cephas got theirs on. "Why the extra power?"

"Don't you understand yet?" Cobalt's voice came over the speaker. "The sphere must be large enough for the Leviathan to fit through. Once every one of the worshippers has perished, the sphere will engulf this entire chamber and the Leviathan will soar through on eagle's wings. And I will assume my rightful place as ruler of the Node of God!"

"Node of God?" Cephas said. "Jonathan, we have to stop this."

"But what about the children?" Renee said.

"Maybe we can get them back." Steel glanced around the chamber. "Maybe we can stop the Leviathan."

"I doubt that." Cephas said. "It's alive."

"Alive?"

"Yes, the other two ships are organic from that region of space. Cobalt constructed the Leviathan using genetic engineering."

"Like the probenosticon?" Steel asked. "I thought it looked familiar." He pulled Cephas around to face him. "Whatever you do, don't let Cobalt go through that interface. I'm going to stop the Leviathan."

"You are going to do what?" Cobalt's voice echoed through the speaker. "I doubt that, Mr. Steel."

"Cephas, the Portal, can you go from that area of space back to this one?"

"Theoretically it should be a two-way connection."

Steel looked down through the gaping hole in the chamber floor. The Leviathan hung in space just meters away from the enlarging sphere. While he still had gravity, he propped his feet against the metal rim at the edge of the glass portion of the floor and banged both fists against the splintering glass. Shards of it broke away and were pulled into the sphere leaving a narrow gap between the enlarging sphere and the floor of the chamber. He grabbed the edge of metal and pulled himself down and out into space. For a second, his head somersaulted through the portal's interface. Dizziness blurred his mind and he was in the other space just inches away from a similar sphere. This sphere was filled with the spectacle of Earth and the crumbling space station. He spun through vacuum and saw the world below. The children's tethered bodies floated meters away. And, approaching them was a black, spiky ship of menacing proportions. The spinning of his body carried him back into the sphere and he was falling onto the surface of the Leviathan.

Nightflies swarmed toward him as he reached out and grabbed the nearest antenna. The surface of the Leviathan was coarse and bumpy like the skin of a man suffering from the worst allergic reaction ever. More nightflies streamed out of a nearby opening. He kicked at the approaching nightflies and spun in the zero gravity toward the open hatch. He grabbed two nightflies by their wings and pulled himself toward them, using their momentum to launch him toward the hatch. He fell inside and was instantly gripped by a gravity field. He landed on his feet and glanced around him. The corridor was grisly and red tinged. He dodged a nightfly and ran around the corridor. There had to be a control room somewhere. He shot down an intersecting corridor and, there ahead of him, was the main control room. It was half the size of one of the smaller ships. Consoles and view screens dominated the floor and in the center of the room, a giant tube pulsed with green and lavender light. The wave of evil it emitted rolled over him like a tsunami. Whatever was powering the Leviathan was in that tube.

He took off a glove and the air was cold but not a vacuum. A nightfly came toward him and he hit it with his fist. The thing screeched and tumbled away in pain. The rest of the nightflies froze and then flew out of the control room. Steel hurried through the consoles and paused before the giant tube. Something within moved and spun and a dark, granular tentacle grazed the glass. His heart raced and he gasped for breath inside the space suit. He hoped he was right about this. He turned and picked up a huge metal chair and smashed it against the tube. The glass cracked. He hit it again and the waves of evil intensified. A tentacle bearing a huge eye appeared from the mist in the tube. One last smash and the glass shattered. Gas gushed out around him and settled on the floor. His bare hand began to sting as if coated in acid.

The tentacle came out of the mist and wrapped itself around his waist. It lifted him off the floor and the thing appeared fully formed. It was the most obscene creature he had ever seen, more gruesome than the probenosticon. Human parts dotted its surface and a giant brain shaped head lowered level with Steel's helmet. One huge green eye blinked.

"What have we heeerreeee?" A voice hissed. "More parts for my final form?"

"No!" Steel screamed. "I am a child of God and I command you back to the hell from which you escaped." Steel shoved his bare hand into the thing's eye. It erupted in blinding red flame and a mouth opened in the top of the brain head emitting an inhuman wail of pain. It dropped Steel and he shoved his hand back into the glove. The thing began to writhe in quick, chaotic movements, smashing consoles and view screens. Steel ran across the floor as the control room began to crumble. Huge hunks of flesh like material fell out of the ceiling. He made it to the corridor. Nightflies scrambled away in panic. He stopped in front of a very ordinary metal hatch. An escape pod! He keyed the access pad and opened the hatch and fell inside. Twelve seats surrounded the interior and he slid into what should have been the control seat. A button blinked before him. "Eject!" He slammed his fist against it and the pod lurched as it was ejected from the ship. The view screen activated and the pod shot away from the Leviathan. It was coming apart at the seams, huge chunks of it falling away into space.

"Jonathan?" Renee's voice said over the speaker. "It's Cephas!"

Chapter 42

Renee watched Steel disappear into the portal and seconds later emerge from its far side to fall toward the Leviathan. Cobalt hung above the sphere and his attention was drawn to Steel's falling body.

"What is he doing?" Cobalt asked.

The artificial gravity was failing but she had enough friction to run across the floor and jump onto the Fallen Throne. The angel's arms were outstretched above it just inches from the beam of energy coursing into the Bloodstone. Even through the space suit she could feel the heat of the stone as it released huge amounts of energy.

She glanced over her shoulder, her vision limited by the helmet. Cobalt was turned away from her. She reached gingerly for the sword and pulled if free of the Fallen Throne. She glanced again over her shoulder and Cobalt was right in front of her.

"What are you doing, Dr. Miller?"

She brought the sword around in one fast swoop and shoved its point into Cobalt's chest. It glowed with blue fire as it sunk through him and out the back of his white jump suit.

"What?" He said. For a second, the lavender fire of his eyes faded and for a second the real dark eyes of Anthony Cobalt appeared. "What happened?" Blood gushed from his mouth and a black pair of wings tore through his jumpsuit, tearing him into two pieces. The faceted, black form of the real tenth demon flew up and away from Cobalt's body as it tumbled into the interface and disappeared.

"He got his wish." The demon said. "He will get to his Promised World after all."

Renee held the sword up before her. "Look around you! You are defeated. Steel is destroying the Leviathan."

The demon's black, faceted face moved as it frowned. "Well, I will still reign over this new world, my dear. And, over your children."

The ring pulsed on her finger. Renee had whispered one word into her ear. She held the ring up. "Recognize this? Kabal! I call you by your true name."

Kabal's wings beat at the thinning air and it floated backwards away from her. "Where did you get that? It is my talisman."

" And, it gives me control over you." Renee stood straighter on the Fallen Throne.

Kabal's face twisted into a sneer. "Control? That is a myth, Dr. Miller. You may have my talisman, but you cannot control me."

"But, I can!" Cephas floated up behind Kabal, his hands extended. The old man grabbed the demon's wings and hugged himself to Kabal. "By the power of the name of the Savior of this universe, I command you to the deepest corners of Tartarus in the name of Jesus Christ!" He screamed.

Kabal bucked against him and Cephas flew across the chamber thudding against a far wall. Kabal spun and his skin cracked. Thousands of tiny crevices covered him and bright, blue light gushed from within. With a silent explosion, his fractured, blackened skin burst away revealing the pale green and lavender creature beneath. The shards of black, jeweled skin whirled and spun and formed a sphere around Kabal. His glowing face filled with fear and he screamed silently into the vacuum now filling the chamber. He tried to dive for the sphere but it contracted into a tiny mirrored ball and disappeared. The Fallen Angel's glow faded and the throne and disc were silent.

The beam of energy continued from the solar satellite and Kabal's wail of pain and torment were cut off by the sudden contracting black skin. It closed in on him and he shoved against the pieces as they squeezed him down into a tiny ball of lavender and green. A crack opened above the empty disc exposing the universe to a ruddy glow from somewhere else and a clawed hand reached through the crack and grabbed Kabal's contracted misery and pulled it down into the crack. It closed with a last gush of flame.

The nightflies disappeared in a flash of dark flames as the demonic spirits were released from Cobalt's creations. Cephas kicked against a space suit stand and sailed through the zero gravity to Renee.

"We did it!" His face glowed with triumph through the faceplate.

"Yes, we did." Renee released the sword and it floated away, droplets of Anthony Cobalt's blood floating in the near vacuum. She looked at the Disc. "But, what about Vega? The children? What will happen to them?"

Cobalt's face slackened and he frowned. "They will need someone to take care of them."

Behind Cephas, the Bloodstone continued to pulse with the never ceasing beam of energy. "The Bloodstone! Can we open the portal again?" Renee asked. "I could take one of the ships. Get Arthur to help me. We could bring them back."

Cephas blinked behind his faceplate and looked away. "They cannot return, Renee. They will die here. You know this."

"But, floating in empty space in space suits with limited air, Cephas. They will die there." Renee pushed away from Cephas and snared the sword. "Maybe if I put the sword back. Maybe there is someone who is still dying in the Worship Sphere."

"Renee, Arthur saved most of them from the energy beam. There were never enough souls to open the portal large enough for the Leviathan."

"But, one of the smaller ships? Yes, we can open the portal again. I can go get them. Arthur can fly the ship." Renee's momentum carried her to the throne. She avoided the beam and slid the hilt of the sword back into the angel's hands. The sword did not glow.

"Someone has to sit on the throne, Renee." Cephas said. "There must be a sacrifice."

"No!" Renee hung in the emptiness of the chamber and her tears leaked away before her eyes.

"Renee." Arthur's voice came over the intercom. "I heard everything you said. Where is Jonathan?"

"He is disabling the Leviathan. Where are you?"

"I just saved Josh. Vivian will take the rest of the survivors back to the surface."

"Vivian?" Cephas asked.

"I know." Arthur said. "I'm just now docking my escape pod with the smaller ship."

"Arthur," Cephas said. "I need you to bring the ship immediately beneath the chamber."

"But, what about Jonathan?"

"He has his hands full. Just do it. Now." Cephas said.

Renee glanced at him in sudden horror. "Cephas, no!"

"Renee, I have lived a full life. It is fitting I give the children a chance. You must take the ship and go after them. You must find a way to keep them alive either here or on that distant world."

Cephas kicked against the floor and he spun towards her snagging the arm of the throne. He pulled himself into the throne and the voice of the Fallen Angel sounded in his ears.

Renee couldn't hear the exchange. Cephas had turned his transmitter off. She touched the toggle switch with her tongue and heard heavy breathing. Was that Jonathan?

"Jonathan! It's Cephas." She screamed into the intercom. The Fallen Throne's eyes began to glow and the throne pulsed with blue light. The arms lifted the sword above the head of the angel.

"Renee? What's happening?"

"Jonathan." Cephas triggered his transmitter. "I don't have much time. I tried to kill Cobalt with the Golem. I used his own evil against him. I must pay the price. There is so much you don't know. Take care of Josh and help him with Arthur."

"No, Cephas. Don't!" She heard Jonathan's voice over

the intercom. The sword lifted high above the throne and beneath the open floor the small ship piloted by Arthur Knight appeared. The ring pulsed on her hand and a foreign voice filled her ears.

"You are the bearer of the Bloodstone, too. I will accept your mastery once the old master leaves this ship."

"Renee, it's Arthur." Another voice came over her intercom. "I can't go with you. The ship will accept only one master and it told me you have a ring with a piece of the Bloodstone. It will only recognize you or me. One of us has to go. It has chosen you. I can't convince it otherwise."

Renee sighed. The inevitability of this hit her hard. "You have to take care of Josh. I understand." More tears clouded the faceplate as she watched in horror. The sword plunged down into Cephas' chest and he screamed in the intercom. Blood gushed from his mouth against the inner surface of the faceplate and the sphere sprang to life from the center of the Disc. With the power of the Bloodstone, it blossomed larger than before until it touched the dome of the smaller ship. It was pulled into the darkness of the far side of the galaxy. Cephas' body tumbled from the throne and through the sphere's interface. Renee swallowed as the quasi-mirrored surface came towards her.

"God, be with me." She prayed and she stepped into the unknown

Chapter 43

Steel grabbed the control stick and the escape pod lurched. "Cephas, don't you do it! Don't you dare!" He screamed. His heart raced and his eyes clouded with tears. "Not now, old man. I need you."

The escape pod responded to his control sluggishly but Steel had no idea how to steer the thing. After being ejected from the Leviathan, it was headed directly toward the ring of the space station. He pulled back on the control stick hoping that might slow down the pod's forward movement. Instead, the pod shot upward and blasted through the crumbling floor of the throne chamber. It rammed the ceiling and bounced back down against the intact portion of the floor. Steel was rattled but realized he was where he wanted to be. He threw open the hatch and floated out into chaos. The chamber was littered with broken glass and shards of metal. The solar satellite beam pulsed against the Bloodstone and he watched Renee Miller disappear into the widening sphere. No sign of Cephas.

He kicked against the escape pod and floated in the direction of the sphere. Through the rippling surface of the interface, Renee floated toward a body tumbling through space. Cephas. He weighed his options. Go after them or stay here. The sphere would eventually collapse and close the Portal. But, now that Cephas was gone and very dead, who would take care of Josh? Arthur! Josh didn't need him anymore. But, Renee and the Children of the Bloodstone would need a defender in their new, harsh world for he had seen the approaching vehicle and felt its evil in the brief moments he passed through that part of the universe.

A figure appeared between him and the sphere. He was tall and clad in green robes with dark, curly hair and jade eyes. "It is not your place to help them." He held up a hand.

Steel paused as if some force gripped him. "Who are you?"

"I am the adversary of Kabal, the tenth demon. I am Donnatto. Events here are not done. Josh will need you. And, there are nine other demons. Will you leave this world to the mind and hands of Lucas and Vivian?"

"I'm sick and tired of these games, Donnatto. Let me pass."

"I'm sorry, but there is grave danger to the world. I can tell you that if that beam breaches this chamber, it will focus on one area on Earth. And, if it reaches Earth, it will ignite billions of sunstones resting in the crater of Diablo Boca. Their power exceeds that of a hundred hydrogen bombs and the crater sits atop a lava field similar to the Yellowstone super volcano. Would you abandon this world and Josh to that fate? It will be like in the days of the dinosaurs when the asteroid stuck. All of life will perish." Donnatto said. "But, this is your choice. You and you alone can make this choice. That is all I can tell you."

Steel gripped his fists in anger. "Why don't you stop it all?"

"This is the doings of man and his choices. God will intervene but God does so through people. He has chosen you as his redemptive vessel. Will you accept this role and save the world?"

Steel's words were cut off by the silent explosion behind Donnatto. The Bloodstone shattered into a million pieces under the onslaught of the power beam. The sphere collapsed in one quick flash of light. Shards of the bloodstone shot away from the explosion like missiles. Donnatto unfurled huge, white wings and blocked most of the shards. But, some of them caught Steel in the chest and the shoulders. The concussion of the explosion drove him back into the escape pod. Pain shot through a dozen spots on his shoulders and his upper chest and one shard of the Bloodstone pierced his faceplate and lodged halfway through. The crystalline sword tumbled end over end toward him. He threw up his arms and the hilt of the sword thudded into his chest. He grabbed the hilt and gasped for breath.

The Fallen Throne broke into pieces and rocketed away into space. The Infinity Disc broke like a stale cookie. The energy beam struck the floor and it began to glow and melt.

"It would seem your choice has been made for you." Donnatto vanished in a flash of light and the entire chamber shook with the aftermath of the explosion. Steel and the escape pod shot through the wall and out into space. The pain was exquisite but it was the hissing of escaping air that caught his attention. Steel was plastered to the surface of the escape pod by his momentum and he tried to turn and find the hatch. His air was growing thin and his sight was fading. How could he save the world if he died in the cold vacuum of space?

Knight spun around in the pod chair. Josh floated in the zero gravity and his eyes were wide.

"It's me, Josh."

"I know. I know." Josh gasped for breath. "I don't know whether to hug you or hit you. You don't know how many times I dreamed of this moment."

"I know you don't understand, Josh."

"I do understand. That's just it. After the past few months with Uncle Cephas and Jonathan, I get it. Sometimes something bigger than you comes along and you have to put your needs aside." Tears formed in his eyes. "But, Vega? Really?"

"Josh, it was a mistake. I was under the influence of a demon." Knight unbuckled from his seat and floated toward him.

"And, this is the part where you tell me you did all for me, right? Faked your death. Abandoned Mom in the middle of her disease. Abandoned me. All because you hooked up with Renee Miller." Josh's face was red with anger.

"Yeah, that's pretty much it, Josh. I screwed up. Big time. We all do." Knight floated closer to him. "I can't expect you to forgive me. Your mother did. That is all I can hope for. I live with the knowledge that I did the best I could to protect you and Vega. I was there watching over you in Lakeside. I was the man in the boat who pulled you from the water."

"You were there? That close and you didn't even talk to me?"

"I was trying to protect you from all of this. Both of us have made huge mistakes but I'm supposed to have known better."

Josh looked away. "You're right. I screwed up, Dad. I let that thing take me. And, when Mom needed me most, I let her down. I let her die."

Knight reached out and took Josh in his arms, pulling him close. "I know. But, your Mom chose to sacrifice herself, Josh. You couldn't have stopped her if you wanted to. It was her choice. And, you and I have to live with the consequences of our choices. But, it's over now. I'm back. All that matters is we stop Cobalt from getting to his Promised World with the children."

Josh's arms finally came around Knight and squeezed him tightly. "I'm sorry, Dad. I'm sorry." He sobbed and Knight's tears mixed with Josh's floating tears.

The pod thumped as it connected with the ship and gravity returned. They landed on their feet just as a voice came over the intercom. Knight wiped at his tears and turned to the view screen.

"Arthur, I need your help." Renee Miller said.

Theo grabbed Josh by the shoulders and lifted him into a huge bear hug. Josh returned the hug. "I thought you were gone, Theo." Josh said.

"Ain't no demon gonna get the best of me." He put him down.

Josh looked around at the interior of the ship's control room. "You can control this ship?"

Knight touched the Bloodstone necklace. "With the Bloodstone. And, now, as much as it hurts me, I have to go."

Josh shook his head. "No, we all go. I just got you back."

"I have to take the ship up to the throne chamber for Renee." His eyes filled with pain. "Renee is going through to the other world to save the children and she will need this ship."

"Are you going, too?" Josh's heart beat faster.

"I want to, but my place is here." He put a hand on Josh's shoulder. "There's something else. Uncle Cephas. There had to be a sacrifice to reopen the Portal. That is why Cobalt was going to kill all of you. Multiple sacrifices to open the portal wide enough for the Leviathan to get through. But, Cobalt is dead. Cephas sent him to hell."

"Bro, that's good!" Josh said. But, there was something painful in his father's eyes. "What is it? Is it Jonathan?"

"Cephas was the last sacrifice to reopen the portal so Renee could save the children." Knight said quietly. "He's gone, Josh."

Josh stumbled back into Theo. The big man trembled. "Papaw's gone?"

"Yes. And, we must honor his sacrifice. I have to hurry and get this ship to Renee before the Portal collapses or Cephas' sacrifice is for nothing." He reached out and grabbed Josh and pulled him close. Josh was numb. "I love you, son. Go with Theo. Take an escape pod and get to Earth safely. I'll come after you in another escape pod. This ship has three more."

"No!" Theo said. "We do this together."

Knight pushed Josh away. He looked up at Theo. "Perhaps you are right. I can fly this ship from the escape pod until it is ejected so let's all get into one."

Josh was reeling, his heart heavy with unexpected tragedy. He had lost his mother. Gained his father. Lost Uncle Cephas. And, where was Jonathan? Was he safe? Theo opened the hatch to an escape pod and motioned in. Josh climbed in and slumped into a seat. Theo followed. Knight buckled Theo into a seat and turned to Josh. He snapped the restraints in place. "You're going to be okay, Josh. I know all about Jonathan. He is a good man with a good heart."

"Why are you telling me this?" Josh jerked against the restraints.

"Because you and Theo are going to be safe."

He hopped through the hatch and slammed it shut. Josh fumbled with his restraints. Theo roared in frustration. But, before he could get the restraint undone, the escape pod lurched as it tore away from the ship.

Chapter 44

"Dude, and people wonder why teenagers don't trust adults!" Josh ripped out of his restraints and floated into the control seat. He sat before the view screen and studied the controls. "I think I can fly this thing, Theo."

"You ought to be able to all those video games you play." Theo fought with his restraints. "What are you going to do?"

"Go find Dad."

"He said for us to go home." Theo said.

"I'm not going anywhere without Dad or Jonathan. We lost Uncle Cephas and that's too much for one day." He said.

"Josh, is that you?" A voice came over the intercom.

"Jonathan?" Josh sat forward. He looked at his keyboard and pressed a button. The view screen switched to a view of the direction from which the radio signal originated. An escape pod moved away from the Eagle's Nest and attached to its side was a figure in a space suit.

"Jonathan, is that you?" Josh asked.

"I'm losing air and I can't get back into the escape pod. Where are you?"

"Theo and I are in a pod." He watched his father's ship moving toward the space station. Who was he going after? "We're coming for you now." He grabbed the control stick and aimed the pod toward the receding figure of Jonathan Steel.

"We have a problem, Theo." He thought furiously. "Jonathan has on a space suit and we will have to open that door to get him inside. But, we don't have space suits."

"That would be a problem." Theo had unfastened his

restraints and was floating right behind Josh. "You got a solution?"

"We can withstand explosive decompression for about thirty seconds. If you can open the door, grab Jonathan, pull him in and shut the door in thirty seconds, I can do an emergency re-compression from here."

"Whoa, little man. How you know this stuff?"

"I saw it in a movie?" Josh shrugged. "Look, I read about it after I saw the movie. I'm a lot smarter than you guys think I am. Besides, I've already survived one decompression today."

Theo nodded. "Okay, so we can do this?"

"You want to save Jonathan?"

"I owe him my life." Theo said.

"Okay." Josh's heart raced as he looked up into Theo's eyes. "But, you have to be fast, Theo. Or, we all die."

"Josh, if you can get me within arms' reach of Jonathan, I can do this."

"Jonathan." Josh touched the intercom switch. "We are almost there. When I get there, grab the hand holds right beside the hatch."

"Josh, what are you planning?"

"Just trust us, Jonathan. Dude, this one time, let us save you."

Josh neared the other escape pod and rotated their pod so the hatch faced Steel. He heard Steel's thump on the outer hull. "Hurry then. I'm losing air." Steel gasped.

Josh punched at the keyboard, speeding through menus until he found what he wanted. He poised with his finger over a bright red button that said 'emergency re-compression'. "Theo, you have to sit in the seat nearest the hatch and buckle in."

"I got this, Josh. Just tell me when to open the door."

Josh flooded his lungs with air and closed his eyes. "Now!"

Theo popped open the door and the air was immediately sucked out of the pod. Josh's head was about to explode and someone tapped his shoulder. He hit the re-compression button and the air returned. His lungs burned and he twisted in the seat. Frost covered Theo and Jonathan floated between them. Josh's fingers were numb from the sudden cold and he fumbled with his restraints. Blood leaked from his nose and formed spheres in the zero gravity.

Theo floated away from his, blood leaking from his nose and ears. Josh pushed him aside and grabbed Steel's helmet. Blood was frozen on the inside of the faceplate. He snapped the helmet off and tossed it aside. Steel's eyes were empty of life and his mouth and nose were covered with blood.

"No!" He screamed. He thumped Steel on the chest and blood sloshed inside. What had happened to him? He grabbed Steel's head and pulled it upward. He opened the man's mouth and started CPR. Steel's blood smeared his face but he blew with all of his might into the man's lungs. He thumped his chest again and Steel coughed, spraying blood over his face.

"Jonathan! You're alive!" He shouted.

Steel's eyes focused. "You're grounded!" He growled. "Where's your father?"

Josh pulled Jonathan up toward the view screen. Behind them, Theo coughed. "Chief, that's one big one you owe me." He said hoarsely.

Steel reached out and grabbed the big man and hugged him. "I get it. We're partners. From now on, you go where I go." He turned to the view screen and grimaced.

"You're hurting." Josh said. "What happened?"

"The big Bloodstone exploded and there are pieces of it in my shoulders and chest. Only superficial, thank goodness. Donnatto protected me."

"Donnatto?"

"An angel. Now, I need to talk to your father."

"He flew the smaller ship to the Portal for Renee." Josh stopped and choked. "Oh my God! Uncle Cephas."

Steel gripped his shoulder. "I know. We don't have time for mourning. We have to stop that energy beam. Once it punches through the Eagle's Nest, it will focus on the crater."

"Those sunstone things?" Theo was shivering from his

brush with frozen death. "There's bunches of those things in that crater."

"Billions, Jonathan." Josh said.

"We have to stop that energy beam." Steel gasped for breath and more blood came from his lips. "I think one of those shards just made it to my lung."

Josh hit the intercom. "Dad, are you there?"

Arthur Knight's face appeared in the view screen. "I'm right here in an escape pod headed toward Vivian's ship. I had to watch Renee go through the portal. She's gone!"

"No time for that! We have to stop that beam. It will hit the crater and release the energy from the Sunstones." Josh said.

"How bad will that be?"

Steel gasped for breath and pulled himself into view. "Extinction level event."

Knight switched the intercom back to Vivian's ship. "Vivian, we have to stop that beam from reaching Earth."

Vivian appeared in the view screen. "And, why should I care?"

Knight swore. "Because if those Sunstones explode it will be like the asteroid that took out the dinosaurs. You wouldn't have any world to play with your magic in."

Vivian nodded. "So, what do you think I can do?"

"Fly your ship up to the satellite and knock it out of orbit."

Vivian frowned. "Well, there's just one little old bitty problem with that. The Barrier."

"What is the Barrier?"

"Have you stopped to wonder why Cobalt didn't just fly off into space to his new world? It's because when the angels fell, God put a barrier around this world to prevent them from tainting the rest of His creation. They can move in and out of the heavenly realms to traverse this world, but this world only. I'm afraid I can't fly this ship past the Barrier."

"I can fly it with the Bloodstone." Knight said.

"Nope. Your smaller ship was different from this ship. This one is too large to be sustained by your puny Bloodstone. You can't fly this without a demon in the cylinder. And a demon can't get past the Barrier."

Knight slumped back in his seat. The smaller ship was gone through the Portal and Renee Miller with it. "What can we do?"

"Well, there is one little bitty thing you can do. Your escape pod has regular manual controls that don't need demonic power. You can fly through the Barrier and take care of that satellite."

Knight sat forward. "Of course. Does it have enough power to get that high?"

"Now, that is the real question, Arthur honey. I guess you'll just have to trust in your faith. There is one thing I can do. I'll fly this ship underneath you and give you a boost up to the level of the Barrier. Once you get through the Barrier, you can use your engines to reach the satellite. Then, just let the escape pod do its job when you fall back to Earth. It will bring you back to Area 613."

Steel gasped for breath and every expansion of his lungs felt like inhaling fire. He struggled toward the view screen. "Arthur, you can't. Come get Josh and Theo and let me go."

Arthur Knight shook his head. "You're bleeding out internally, Steel. Josh needs to get you to Earth and to a hospital. I can handle this. All I have to do is nudge the satellite and the beam will swing away from the crater."

Josh touched controls on the console. "How much reserve fuel do you have, Dad?"

Knight glanced down at his console. "85%."

"Okay, let Vivian give you a boost with the ship and let the momentum carry you past this Barrier, Dad. Don't use the guiding rockets until you're almost to the satellite. You'll get more bang for your buck once you are closer to its orbit."

"You have it all figured out, don't you?" Knight said. "Son, I'm sorry I never told you how proud I was of you. You are a genius, you know."

"Yeah, tell that to Jonathan." Josh laughed quietly.

Theo tapped the small port in the hatch. "Chief, that beam is tearing the space station to bits. We gotta hurry."

Josh switched the view from his father to an external camera. The Eagle's Nest had torn itself into multiple huge fragments. Most were plunging through the atmosphere in a fiery display of destruction. The beam was still focused by the worship sphere and was still contained by the blossoming debris of the station.

"Josh, get Jonathan to safety. I'll take care of the satellite."

Josh nodded and his finger poised over the execute button. "Dad, I love you."

"I love you too, son."

Josh pushed the button that would take them back to Earth.

Vivian's ship hit the pod with a terrifying jolt and Arthur was thrown back in his control seat. His view screen showed the gleaming copper colored flower of the solar satellite. The gravitational force increased and his bones and flesh pressed into the seat. The escape pod creaked with the increased pressure and then, just as suddenly as it had come, the force was gone and he was hurtling toward the satellite. For a moment something passed over him, a wave of power, the Barrier. Then, he was sailing through space toward the satellite.

The alarm caught him off guard. He glanced at a readout. His engine power had fallen to 15%! What?

"This escape pod was never meant to undergo those kinds of gravitational forces." Someone said.

He glanced over his shoulder. A man dressed in green robes hovered in the escape pod. "What?"

"Donnatto. Cobalt's adversary. I'm a messenger from

God, Arthur. And, yes, we can pass through the Barrier with ease. I'm here to help you."

The alarm blared away and the readout dropped to zero. His fuel had all leaked away. "That is not good, is it?"

"No. Without fuel you cannot reach the satellite safely. Nor, can you return through the atmosphere without burning up."

Knight gritted his teeth. "Then, why are you here? What can you do for me?"

"I can encourage you. I can be with you so you will not die alone."

Knight blinked furiously and nodded. "I should have expected this. I made a lot of mistakes. Why would God give me a break now?"

"You have made many choices, Arthur Knight. Many of them were good. Some were bad. It is your choice that has brought you here. Every choice brings you to a point of faith. Where is your faith, Arthur Knight?"

Knight glanced at Donnatto. "It has been tested, angel. Tested and found wanting at times. But, right now, I have to concentrate on saving my son. If I don't stop that satellite, Earth is doomed."

"Precisely. It is your love for your son that is a sign of your faith."

Knight watched the satellite grow closer. He called up a graphic on his view screen. He would miss the satellite's main body by ten feet!

"I'm going to miss the satellite. Hurtle out into space where I can have a ringside seat of the Earth's destruction."

Donnatto hovered closer to him and gently touched his chest with a brush of his hand. "You have something else with great power."

Knight glanced down. The Bloodstone! "But, this stone was always powered by evil."

"Since the day you placed this stone around your neck again, I have been the power behind your Bloodstone." Donnatto smiled.

"Then I can use it like I did in the cave. If I can concentrate a burst of energy to the body of the satellite, it might nudge it out of synchronous orbit."

"It is your choice, of course." Donnatto said.

Knight turned to the small port in the hatch. "I'll have to have it in my line of sight. And, the beam will pierce the port window. I'll lose all of my air."

"Yes."

Knight turned back to his console. He had only seconds before zooming past the satellite. He studied the console and hit a record button and rattled off a quick message. He sent it out and then unbuckled from his seat. He pulled the bloodstone from his shirt.

Donnato appeared beside him. "Just concentrate on sending a powerful burst of energy at the satellite."

"Maybe I can hit that instrument panel on the side of the satellite facing us. The burst will not be powerful enough to take out the mirror. But, with the instrument panel disabled, the satellite should spiral out of control." Knight said. His heart raced and he drew a deep breath. He imagined a ruby red energy beam slamming across space into the instrument panel. He cleared his mind of guilt and fear and began to pray for the power of concentration.

"I've made a lot of mistakes, Donnato. I hope that God can forgive me."

"You cannot imagine what it is like to be forgiven, Arthur Knight. We chose early and those who chose wisely never needed the forgiveness of our Lord. Those who rebelled placed themselves forever beyond forgiveness. You have been given a special gift, Arthur Knight. You have been given grace."

Knight blinked away tears. "I have been forgiven. I don't deserve it." And with that thought, he found the power. The Bloodstone glowed with energy and warmed as it floated just inches from his throat. Pressure built within his mind as if a dam were about to burst. He concentrated on the approaching satellite and as the pod passed beneath the instrument panel, a red hot light exploded from the Bloodstone. A beam of coherent light shot forward, piercing the window and crossing the black space between the pod and the satellite. The light burst against the instrumental panel in a shower of red energy. The satellite shuddered. Waves passed across the underside of the huge parabolic mirror and the satellite began to spin slowly.

Air hissed around Knight. The pod had now passed beyond the satellite and began to slowly rotate from the impetus of the energy beam. The Bloodstone floated dark and lifeless before his eyes as the window turned toward the satellite. The massive mirror undulated with the torque from the satellite. The beam was swinging slowly and inexorably toward the escape pod. In just seconds, he would be fried. He blinked and searched the cabin for Donnato. The angel was gone. Outside the window, the sparkling red energy beam swung closer.

Arthur Knight closed his eyes and put his arms out to the side. He was floating in zero gravity; as he had soared high above the earth the day he chose to die. The wind fluttered against the fabric of his hang glider. In his memories, he glanced down as the hang glider slipped off the edge of the cliff. Below him on a lower ledge his son smiled at him. Josh was only eleven and his face held all the promise and hope of a young life. Arthur knew the days ahead would be tough for Josh and Claire. They would search for his body but he knew his sacrifice would save their lives. They were destined to die because of his bad choices of the last few years. But, this one choice would guarantee they would live. Josh flashed beneath him and he glanced over his shoulder one last time to see the boy standing on the cliff edge as his mother put her arms protectively around him. He looked forward again toward the red setting sun, toward the uncertain future with the painful but certain knowledge that he was doing the right thing.

The energy beam bathed the pod and the window ruptured. The collapsing window allowed the air to be sucked outward and Arthur Knight flew out of the escape pod. His arms wide, he soared through the fractured window into the dark night and into his eternal future.

Chapter 45

The escape pod shook with the descent through the atmosphere. Occasionally, the rockets would fire and adjust its path. Steel glanced at Theo strapped into his seat. Josh was behind the control panel. Every breath was a fire in his chest.

"What's happening with your father?" He whispered.

"I don't know. Dude, I've lost track of his pod. We're in the atmosphere right now. Heat is building up around us and keeping us from receiving anything." Josh screamed above the buffeting wind.

"You're gonna be fine, Chief." Theo said.

Gravity returned as they left the heated thickness of the upper atmosphere. The pod continued to fall and Josh pointed to the console screen. "Two minutes and we'll have a parachute descent. Looks like we'll land in Area 613."

Something beeped on the console and Josh touched a key. "We're receiving a message. It's from Dad."

The screen lit up with Knight's face and behind him, the face of Donnatto was visible. "Who's that?" Josh asked.

"Donnatto. An angel." Steel said.

"Josh, I hate to tell you this but I just found out this is a one-way mission. When Vivian boosted the ship, I lost my fuel. I'm going to take out the satellite with my bloodstone but the pod will continue on out into space."

"No!" Josh screamed. "I just found you again!"

"Listen carefully. Jonathan will take care of you now. He's all you've got left. He's a good man. He was there when I first saved some of the children. I saw him before he lost his memory and he was as good then as he is now. Understand?

Good. Because I love you more than you can ever know. I'm doing this for you and your future. Remember that. Just like your mother did."

Josh slumped back in the seat as the message continued. "And, Steel, I know you. I know everything. I don't have time to tell you. Just remember this one thing. No matter what anyone says, you are the good one. Understand?"

"The good one?" Steel said. "Are you talking about my father?"

"About your father." Knight said as if he anticipated Steel's thoughts. "Your father is very evil. But, deep inside he has a core of goodness. What he has done to you he did for a greater good but somewhere along the way he got lost in the darkness. Remember that when you finally find out the truth. Take care of my son. Promise me."

"No! This can't happen this way. Let me take the pod." Steel said.

"Too late." Josh whispered. "It has already happened. This is a recording."

"Promise me, Jonathan." Knight's image faded on the view screen.

"I promise." Steel said.

The pod lurched with the deployment of the parachute. The view screen had switched to a default image of the outside of the ship. Chunks of the falling Eagle's Nest filled the air around them with flaming metal. Below them, the desert burned and smoked with fire. As they neared the crater, they saw the crumpled remains of the sunstone production apparatus. Only a small amount of smoke and fire burned in the depths of the crater. With the cessation of the energy beam, the world was safe, thanks to Arthur Knight.

The escape pod landed outside the crater. When the hatch to the escape pod opened, soldiers with formidable guns appeared backed by none other than FBI Special Agent Franklin Ross.

"Hold up, guys. As much as I'd like to see you pull those triggers these are the good guys."

Theo popped out of his seat. "Ross, we need evacuation now. Steel is hurt bad."

"And, Cobalt?" Ross motioned to a nearby vehicle.

"Dead. Gone for good."

Behind them flaming metal rained down from the sky and fell into the empty desert. "I figured as much when his precious nest began to fall out of the sky. We found Dr. Barnard and managed to disperse the sunstones out over the desert where they exploded harmlessly."

A team of medics appeared at the door and took Steel out of his seat. He squinted into the bright sunlight and watched in the far distance as a saucer shaped craft settled into the compound on the edge of the crater. "Theo, take care of Josh." He whispered.

Vivian ushered the survivors out of the ship and onto the stage they had left behind just hours before. She counted eighty-five of them. Most were fine but a few had burns and their clothes were in tatters. They gathered at the edge of the stage and looked out over the smoking remnants of Diablo Boca. Vivian took her place before them and fought the overwhelming weakness in the absence of her demons.

"My fellow worshippers." She began. As much as she hated what she was about to say, she had promised Bile he would have new followers. "You have survived a terrible tragedy but you must know that you were betrayed by Anthony Cobalt. He killed Magan Celeste and was planning on sacrificing all of you."

Murmurs came from the crowd. "But, thanks to a spirit guide sent to us by Magan Celeste, some of you survived to carry on her work."

Most of them seemed unaffected by her words. "Behold you new leader." She motioned to the ramp leading down from the ship. Bile appeared tentatively at the top of the ramp and glanced out. The spiral tattoo was missing from his eye.

"Uh, Miss Vivian, I am not the chosen one." He stepped out onto the ramp and some of the survivors began to mumble. "Wait! The true leader, the true spirit guide has appeared within the ship and wishes to address everyone."

Bile hurried to Vivian's side and she glared at him. "This is not what we agreed on."

"There's been a change of plans." Bile said.

The entryway at the top of the ramp brightened with lavender light and a tall, white skinned figure stepped into view. He wore a full length black coat that brushed the ramp surface. His white chest was bare allowing everyone to see the tattoos that moved and danced over his skin. His bare scalp glistened in the stage lights and he smiled, revealing huge white teeth.

"Welcome my children. I have been planning on this for so long. I am Lucas and soon, you will worship me." He said.

Vivian cursed and headed for the ramp when an iron hard grip on her arm stopped her. She looked up into the eyes of Bile, now sporting the spiral tattoo. "This was his plan, Vivian. The plan all along."

She jerked out of his grasp and glared at Lucas. He ignored her as he descended the ramp and stood on the edge of the stage. He shrugged off his long coat and it fell to his feet. His bare abdomen and chest glowed with light. "Now, my children, if you will watch closely, there are many totems alive on my skin. They have journeyed far to find a host with which to bind in harmony. If you will worship me," he gestured at one of the burn victims. His blistered skin disappeared and was replaced with new, shiny skin. "I will heal your wounds and you may host a spirit guide from beyond the stars. All you have to do is invite me in."

They fell as one at the feet of Lucas. Bile smiled. "There's a sucker born every day!"

"What about my demons?"

Bile cocked his head and shrugged. "I have them all safely tucked inside. Would you like them back?"

"Yes!" Vivian hissed.

Chapter 46

Vivian Darbonne Ketrick Wulf wore a full length shimmering black evening gown. She paused outside the entrance to the ball room in the macabre castle off "Maimed Street".

"How do I look?" She asked her demons.

"Divine." Vitreo whispered. It giggled. "Pardon the pun."

"I am finding it difficult to work with your demons." Reyjacklik said in her mind.

Vivian straightened her hair and took a small compact mirror from her small purse. "Rey, baby, you align yourself with Summer and my shark demon. Vitreo, I am in charge. Remember that. If you don't like your current situation, then get out of my head. But, if I were you, I'd buckle up because what I'm about to do will make Council history." She studied her reflection in the mirror. For a second, her demons appeared in the depths of her eyes. The Vitreo demon caused her eyes to turn ghostly white.

"I have always been and shall always be your servant." Vitreo said. Her eyes returned to normal.

"Just remember that I beat the Master's second. I defeated Kabal." She snapped the compact closed and ripped aside the heavy curtain separating her from the ball room. The Council Members were having one of their feasts. The last time, she had been forbidden from staying and as she watched the zombie like servants whisk away the remnants of food, she was glad she had not participated. When she was in charge of the Council, the feasts would be catered.

Without hesitation, she sauntered into the room and walked straight through the table allowing her demons to reconfigure the molecules as she moved through the ancient wood. She shoved aside serving carts loaded with unspeakable organs of unknown origin. She swooped up a decanter of red liquid and gulped it down. It burned her throat and brought tears to her eyes. She stopped in the center of the circle and slowly turned to survey the assembled Council of Darkness.

"I did as you said. I demand my seat on the council."

Mumbling and murmurs came from behind the veils. Some were filled with outrage. Cursing in languages she would never understand assaulted her. She tossed the decanter aside and it shattered on the stone floor. "Quiet!" She screamed.

The council members fell silent except for one high pitched voice that giggled. "Oh, you are such a drama queen. If you did not amuse us, we would disembowel you where you stand and have you for dessert."

The raspy voice of the Council Master came from behind her. "Enough! She has broken protocol but I decree we give her a moment of tolerance. Vivian, you did not prevent the children from going through the portal."

Vivian whirled and tried to pierce the veils with her gaze. "I was told to keep Kabal from going through the portal. And, I was told not to kill Jonathan Steel. I have kept my end of the bargain. Kabal is in Tartarus. Now, you will keep your end of the bargain."

"Just a moment." A voice came from the entryway.

Vivian turned and watched Lucas materialize outside the table's perimeter. "If it had not have been for my intervention, Vivian would have failed."

Vivian clenched her fists in anger. "Lucas! You showed up at the last moment and claimed the allegiance of the remaining Enochians. I hardly call that intervention."

Lucas walked toward her and gestured to the side table. A yellow bolt of energy came from his hand and hit the egg shaped Grimvox. It sizzled and slammed shut. The young woman fell back away from the closed Grimvox and seemed to awaken from her trance. The Keeper, head bandaged with bloody gauze from his last encounter with the Council kept her from falling out of the chair.

"I protest!" He shouted.

"Silence, Keeper." Lucas said. "I invoke the Privacy Clause reserved only to the right hand of the Master. This is not to be recorded."

The Keeper frowned and nodded. "I recognize your rights."

Lucas nodded and stepped through the table substance. He wore a red, full length coat over his dark shirt and pants. His eyes glowed in the torchlight. "Now, let us get the facts straight, shall we? I was the one who began this entire process, Vivian. I saw that the Bloodstone was in play to produce the Nephilim. I recruited Arthur Knight. I found the twelve parents willing to sire the Children of the Bloodstone. I assisted Kabal at every stage of the way. You just happened to get in the way at the last moment."

"And, your plan all along was to acquire the church of the Enochians and eliminate Kabal. I was the one who took out Kabal and I could care less about the rest of your plans. My presence here is to claim my reward because I did as the Council instructed."

"Did you tell them Cobalt knew your plans all along?" Lucas paced around her and ran his hand over his bare scalp. "Did you tell them you were in league with his plans? Did you tell them how weak you became without your demons and how you saved the Enochian worshippers from certain death?" He drew closer to her until the red veins in the white of his eyes could be seen. "Did you tell them how you turned one of Cobalt's demons to your side?"

"What does it matter how I achieved these results? Kabal is no longer among us and did not go through the Portal. That was the task I was assigned." Vivian said.

"Well, Vivian, you are more naïve than I thought." Lucas turned slowly and studied the chairs of the members of the Council. "I have long been the right hand man of the Master. Many of you I recruited personally. All of you know I have the Master's ear. With his busy schedule, he has relied on me to keep him appraised of your progress as a Council." He returned his gaze to Vivian. "I have kept my silence regarding many of your secrets. I have never divulged them to the Master. But, now, I am cashing in my chips. I demand that this vixen be banished and her demons harvested. I demand that she be utterly destroyed. And, I demand to become Master of the Council."

Vivian paled and stepped away from Lucas. "What?"

"How dare you make such demands of the Council." The woman with the Scottish accent said.

"Even I am not this brazen, Lucas." Vivian said.

Lucas' large, white teeth flashed. "Don't you see? Are you such fools? Every member of this Council has known of the portal our Master has sought for millennia. Everyone one of you knows how important this portal is to the Master but you desired it so you could go through this Portal to the Promised World. You knew Kabal was looking for it and you kept that a secret from the Master." He smiled. "And, the Master does not favor those who keep secrets from him, does he? To make matters worse, if not for Jonathan Steel and his meddling humans, the Portal would have remained open long enough for each of you to pass over. You thought the Portal would be opened at Diablo Boca, a quick teleportation across the country. You did not anticipate Vivian rescuing the Enochians and preventing their death from keeping the portal open for you! And, you did not anticipate the Portal would be open just kilometers from the Barrier. I have much to tell the Master about what you have not told him, it would seem. Unless this Council meets my demands."

The Council members broke out in protests. Angry outbursts and arguments began. Vivian stormed over to Lucas.

"No!" Vivian hissed. "You cannot blackmail this Council like this. When the Master finds out--"

"The Master is far too busy to waste his time with the Council." Lucas smiled.

"Am I?" A voice filled the chamber and brought dust falling from the roof. Vivian clamped her hands over her ears as the voice thundered around her. "I will speak."

A ruddy light burst down on them from the heights of the ceiling. Vivian looked up into the cone of harsh light and watched a figure descend out of the shadows. She stepped back until the edge of the table cut into her back. She watched Lucas' face fill with fear. He fell prostrate on the floor of the chamber. In every niche around the table, she saw the shadowy figures kneel before their coming master.

His presence descended upon Vivian like a huge, dark smothering cloud of hatred and evil. Her demons rejoiced within her. They shrieked with delight and for a moment she was herself. In that moment she felt so much regret, so much remorse, so much guilt that her heart sank with the weight of it. She slumped down to the floor and leaned back against the ancient wood of the table. Her eyes were riveted to the descending figure. She could not tear her gaze away. He was ghastly. And, he was beautiful.

The Master appeared as a man in his mid forties. He had short, dark hair shot with streaks of gray. His face was ageless with an aristocratic nose and high cheekbones. He wore a white suit. His eyes glowed with the pale blue of the ice from a crumbling glacier. He settled effortlessly to the floor in the center of the room. The reddish light vanished and he was an ordinary man standing over Lucas.

"You foul thing! Stand up!" His voice thundered around the room and blood trickled from Vivian's ears.

Lucas slowly stood to his feet. He shook with fear and his eyes were averted to the floor. "Master, I beg your mercy."

The man crossed his arms and laughed. It was a cruel, capricious laughter that carried with it no mirth, only derision. "You know I hate mercy, Lucas. You have disappointed me again. Did you think I was unaware of your work with Kabal? Did you think I would not know Kabal was seeking the components to once again open the Portal?"

"Sir, I was doing the work of the Council."

"Liar!" One of the members shouted.

"Silence!" Thunder shook the building. The Master turned slowly to survey the Council. "You sent this woman to stop Kabal from opening the Portal. You knew about the

Portal? What was the plan of this Council? To flee from my rule and seek a place of power in a far land?"

Utter and complete silence ruled until the Master's face darkened and he shouted. "Answer me!" The remaining windows in the great dining hall shattered. The storm raging outside burst into the room with wind and rain. Broken bits of stained glass rained down on them. The man slowly turned and gazed at the members of the Council and finally, he looked down upon Lucas. "You cannot lie to me, Lucas. I ask one more time. Why wasn't I informed of the opening of the portal?"

"We did not know you were interested, Master." Came a woman's voice. Whereas it had been powerful before, now it was weak.

Vivian found some strength and extended a finger to point at Lucas. "He knew. Master, he planned it all along. He denied you access to your rightful place among the stars. He worked with Kabal to find the Bloodstone. His plans have taken years. Planning implies premeditation." Vivian stood up and straightened her dress. "Lucas alone has denied you these things. You cannot blame the Council. He told them he has your ear and they thought you knew all along. They did not know the ancient story of how you were denied access to the world that shines in the deep night sky like a jewel. But, Lucas knew this and he chose to help Kabal instead of his Master."

The Master stepped over Lucas' prostrate form and drew closer to Vivian. Her demons writhed in panic. He towered over her. "Vivian, my dear, Vivian. You brought me such pleasure during our evenings on the town." Vivian gasped. While searching for the Ark of Chaos, she had drawn close to a deputy sheriff. And, in a dream, he had become the Master incarnate.

"I told you then to return to your efforts. And, you have. You alone this night deserve my mercy." He spun and glared at Lucas. "However," His face twisted in anger and he flicked his wrist. Lucas floated upward. The man casually flicked his other hand and Lucas began shrieking in agony. Vivian wanted to put her hands to her ears but she was paralyzed by the sight of Lucas writhing in mid air. His clothes split in great tears and blood splattered from within. The Master opened his left fist. Lucas' tattoos writhed in pain on his body.

The tattoos leaped from his skin leaving behind the bloody outline of their shapes. The Master motioned above his head and the tattoos swirled into a ball of shadowy mist. Within the mist, their light flared crimson as each tattoo assumed the shape of its demon. Over a dozen demons screamed in silent agony as the mist formed into a crystalline ball trapping them within. The Master gestured with his left hand. A black pedestal appeared in the center of the table's inner circle. The crystal ball came to rest on the pedestal. Within, Lucas' demons pressed against the transparent surface. Their eyes and teeth and mouths gyrated as the demons teemed within.

"I take your possessions." The Master said. "You are powerless from now on."

"No, please Master!" Lucas pleaded. The bloody outlines of his tattoos leaked blood over his white skin. The Master snapped his fingers and Lucas' body soared to the apex of the chamber into the shadows. His shrieking continued and blood rained down upon them in a gentle shower. Vivian felt it on her cheek and saw it splatter in great drops on the grimy floor. None of them touched the Master. And then, he turned his gaze on her. His eyes glowed with energy and power.

"My dear Vivian. I cannot give you a seat on this Council. They have already chosen the next apprentices. I know this disappoints you but let the lesson of Lucas sink deeply into your mind." He turned and slowly regarded each seat around the table. "And, this message is also for each of you. The plans you carry out are MY plans, not yours. If you forget that again, your pain and suffering will be ten times greater than that of Lucas."

In unison, the voices said, "We understand, Master."

The Master gestured to the Keeper of the Grimvox. "And you shall never seal the Grimvox, Keeper. I revoke any and all claims to privacy from this moment on." The Keeper smiled.

"Of course, Master." The Grimvox reopened and its eerie

yellow glow played over the features of the young bald woman. She fell once again into her trance.

The Master turned to the members of the Council. "You will no longer keep any secrets from me, do you understand?"

In unison the members said, "Yes, Master."

"To insure your co-operation, this crystal ball filled with Lucas' demons will remain at the center of every meeting of the Council." The Master nodded and took Vivian's right hand. He reached down with his generous lips, his maroon eyes glowing and touch them to Vivian's hand. The kiss was hideous and filled with a mixture of agony and ecstasy. Vivian convulsed and then fought for control. "I have given you the Master's kiss." He whispered. "It has been used once before on the Son. Do not forget the importance of this gesture, Vivian."

"Thank you, Master." She whispered. "I will not."

The Master snapped his fingers and Lucas' body plummeted from the ceiling. It smacked into the floor in a splash of blood and gore. His skin was covered with a thousand rips. Lucas' chest still moved and the Master bent over and whispered something in his ear. Lucas' eyes flew open and he tried to shout but all that came from his mouth was gurgling. The Master stood up and slowly turned to survey his Council.

"Your decisions have cost me much this day. Remember that. Come, Vivian. I will walk you to your vehicle."

A shudder ran down her spine. The demons within her wriggled in a mixture of fear and delight at the sound of Lucas' ragged breathing. She fell into step beside the Prince of Darkness as they moved through the substance of the table and entered the tunnel leading outside.

"What will happen to Lucas?" She managed in a hoarse voice.

"He has always been my tool, Vivian. He will recover. But, he becomes, at times, too arrogant and too independent." His voice was dripping honey as he touched her arm and paused. He looked down at her with wide eyes that reflected the flickering torches in the hallway. "All of my servants must be humble, Vivian. All of them."

"I understand." She whispered.

"You have shown a great deal of incentive and inventiveness. Vivian, my dear, the Council of Darkness has been the same for centuries. Each member's demon has found new hosts as my plans have unfolded throughout the years. But, lately, they have grown complacent. In today's world, humans are so easily duped and so easily drawn into temptation. It is, how do you say, like shooting ducks in a pond?"

Vivian swallowed. "Something like that."

"I must confess some on the Council are too cautious. Others, too bold. I knew the Creator would grow overconfident one day. My victories over the past two centuries have been considerable." He leaned into her and his breath was cold as it bathed her face. "And, I am gaining momentum. I can see my victory ahead. This is why I no longer need the Promised World. Earth is my Promised World. Now, Vivian, what you have offered me is a fresh new start. You bring new blood to the struggle."

"So, one day I may have a seat on the Council?" Did she dare hope?

The Master laughed and placed a hand on her shoulder. Her demons recoiled in fear at the touch but she did not flinch. "A seat? One day, you will BE the Council. But, until that day, I have a task for you."

"Whatever my Master desires." She bowed her head slightly.

"Vivian, you have shown great ambition to ascend to the supreme position of the head of the council and in that desire, you have bested four of my most powerful demons."

She looked back up and into the fiery eyes. "Then, I am to be punished?"

"No, my dear. I want you to continue. Each of these demons had deviated from the master plan into areas that would benefit themselves. I will not tolerate such lack of loyalty. I have a master plan, indeed, and it has been in place for almost two thousand years. If I dissolve the Council, it will set me back by centuries. You have shown the cracks in that plan. You have revealed the weakness in my own Council of

Darkness." The Master clasped his hands in front of him and moved down the tunnel. Vivian followed

"What would you have me do?" She asked as she caught up with him.

"Continue with your plan to destabilize the Council. Root out those whose plans no longer comport with mine." His hair was tumbled by a sudden gust of humid air from the open end of the tunnel. Torches guttered around them and flared out into total darkness. In the complete darkness, the only light came from the Master's pale, blue eyes. "Weed out the dissenters, Vivian. Test them. Task them. If they have deviated from my plan, I will know it and you will lead them to defeat even as you gain more strength. And, if they are fulfilling my plan, not theirs, then you will be rewarded for helping them."

"Then, I can continue. With your blessing?"

"Oh, the council can never know, Vivian. I will deny we had this conversation. If your plan suffers too much, then I will have to destroy you." He leaned closer and the glare of his glowing eyes played across her face. "But, I have faith in you, Vivian. You are a child of the twenty first century and it is in this time I will defeat the Creator."

"And, what of Jonathan Steel?"

The Master straightened, and for an instant, the glow in his eyes weakened. But, they flared with anger and confidence. "The man you know as Jonathan Steel has been given to me for a season. The Creator has allowed me a free hand to deal with him as I see fit."

"Like Job?" Vivian asked.

The Master was silent for a moment. He looked away. "That seemed a good idea at the beginning. But, when I am finished with this man, the ending will be far different. If you must, let Jonathan Steel continue to think he is winning the battle. It is the war I am concerned with. Unlike Lucas, I trust you will be more subtle."

Tears filled her eyes and she nodded. "I will continue my own plan, Master."

She watched his hand flutter in the darkness and the

torches flared to life again with glowing flames. "Yes, you will continue, Vivian. You have accumulated much power and wealth and these are the tools of my kingdom. Use them well. I will send one member of the Council to direct your path. Cooperate." He leaned forward and heat emanated from his eyes. "Or, your fate will be far worse than Lucas'."

Vivian stepped back and her Master straightened. His laughter echoed down the tunnel as he faded from sight, swirling in a cloud of red light and mist. She walked to the end of the tunnel. Bile waited beside the limousine. The night began to clear as the storm receded. She studied Bile's face, the pulsing tattoo around his right eye.

"I take you heard all of that?"

Bile was speechless for a moment. "It would seem you have the upper hand for the moment."

"I agree." She slapped him hard across the face. The tattoo pulsed and fury filled the man's eyes. He swallowed and rubbed the red spot on his cheek. "Now, Bile and number thirteen, from now on, you do as *I* say." Vivian hissed.

"You are mistaken, sweetie." Someone said behind her. She whirled in surprise. A figure paused and a match flared. He held the match up to the bowl of a Meerschaum pipe and for a second the flaring flame illuminated his Panama Hat and his turquoise eyes. "The Master instructed me to continue with a plan I started twenty years ago. Together, we will bring down the Council."

"And, just who are you?"

"The Council knows me as the first demon. But, you can call me the Captain."

Chapter 47

Steel opened the back door of the condo. It had been five days since the fall of the Eagle's Nest. Most of the pain had been relieved when the surgeon had removed the pieces of Bloodstone from his chest and arms. Cassandra Sebastian stood on his doorstep.

"Long time no see." She said.

"You made it back from Axum."

"No thanks to you." She dropped the smile. "Okay, so I double crossed you."

"You're a consummate liar, Cassie." Steel said. "I don't trust you any further than I could throw you and right now, I'm pretty sore."

"So I heard." She looked away and squinted in the bright sun. She touched the corner of her eye. "Renee is gone then."

"Yes."

"When she called I told her I was sorry. I asked her to forgive me. She hung up on me." She dabbed at her eyes and looked away. "I will miss her. For real."

"Why are you here?"

"I heard today was Josh's birthday. Saw it on Facebook. I have something for him." She gestured to her truck. "I'm here to clear out Renee's stuff from her condo in Pensacola and I had some of my people in Europe looking for something." She motioned to the back of the truck as Josh appeared beside Steel.

"Dude! Is that my motorcycle?"

Sebastian smiled. "We found it in Romania."

Josh ran past her and Theo stepped out onto the porch. "Hey, little man, let me help you with that."

Theo lifted the motorcycle from the back of the truck and sat it on the ground. Josh beamed as he ran his hands over the seat. "It's good as new."

"I had it cleaned up and tuned for you." Sebastian said.

"What's the catch?" Steel asked.

"You hurt me, Steel." Sebastian pouted. "I'm bringing a gift for Josh. That's it."

"No, I know you."

"Okay, so I'm trying to turn over a new leaf. Renee is gone. My show has been put on hiatus. Advertisers are suing. I'll probably have to declare bankruptcy. I was hoping to tell Renee's story. You know, to honor her." She glanced away and wiped at her eye. "I'm not the monster you think I am."

"I never thought you were a monster. Just a narcissist." Steel said.

Sebastian glared at him. "I deserve that. So, tell me. Did you see the Ark?"

"I thought that would come up. Yes."

She paled and wiped at her lips. "Tell me about it. Did you take pictures? Will you be a guest on my show once I get it going again?"

"I can't." Steel stepped back inside the kitchen and grabbed the canvas bag from the corner. "But, I have something for you only if you promise me you will tell Renee's story. And, Arthur's too."

She looked the bag with hungry eyes. "What's the catch?"

"Tell their story. Not yours."

Sebastian's eyes widened and she reached for the sack. Steel pulled it away and reached inside. He withdrew the crystal sword.

"No way!" She screamed as he held the translucent blue sword up to catch the sunlight. "This is perfect! This will save my show, Jonathan." She reached for the bag again and Steel slid the sword back inside.

"First, you tell their story. Then, you get the sword."

Sebastian nodded and drew a deep breath. "You know that sword could possibly save my life."

"Yes. The monk told me a prophecy. We will stand on the place of the skull, whatever that means, Cassie. And, you will be healed."

Cassie planted both hands on her hips and glanced off into the distance. "You would stand beside me?"

"Looks that way. But, until then, a lot has to change. You have to change."

"I will." She turned back to him. "I'll tell the story not only for Renee but for the children, too."

Josh pushed the motorcycle to the foot of the back porch stairs. "Vega was my half sister. Don't mess this up."

Cassie nodded and climbed into her truck. "I'll try not to. I'll do what's right. I promise."

Would Cassie keep her promise? Was there hope for someone as lost as she? There had been hope for himself. It was a comforting sight, one that Steel sorely needed. His flashback had taught him something powerful. He had connections in this world that gave him strength and purpose. Vega had connections with her "invisible" friends and Donnatto had come through in the end delivering all of them from certain death. He could only hope that the children were safe in God's hands. It was also comforting to know he had connections with total strangers who agreed that his father was a horrid creature. There was Major Miller and the stranger, Stoneheart. Steel was not alone in his pursuit of the man called the Captain.

Back inside, Theo and Josh sat around the birthday cake. The candles glittered with flashes and Theo sang out loud and clear. Josh blew out the candles in a powerful puff.

"I always knew you were full of hot air." Steel said. He walked out onto the deck and sat down in a beach chair overlooking the ocean. Two small children were playing at the edge of the water. Their sand castle was imperfect and crooked but they were proud. They just could not see the slowly rising tide of the ocean. Within an hour, the relentless sea would take their creation from them.

"Thanks for the party." Josh appeared at his side and sat down in the other chair. "I wish Uncle Cephas could be here."

"And, your father, too. Happy Belated Birthday. Number seventeen, right?"

Josh nodded as he placed a plate with a piece of cake in front of himself and another plate in front of Jonathan. "Bro, it looks like good cake. Theo is a good cook, isn't he?"

"Yep." Steel said.

Josh took a bite of his cake. "Dude, Theo made me a cake and now I have my Dad's motorcycle. I'm going to miss him. Again. Theo said there were some more of my Dad's cases in storage. Think we should look through them?"

A cold chill coursed down his spine. Knight had known Steel in his former life. Was there evidence in that storage shed? Did he really want to learn those secrets? "Not today, Josh. Today is your day." He said. Steel reached into his tee shirt pocket and took out the small pouch. "I didn't know what to get you for your birthday but I thought you might like this."

Josh's eyes widened. "No way! You got me something?"

"Well, we have been kind of busy but I managed to find a gift. But, before you open the pouch I need to tell you something."

He watched Josh's face cloud up waiting for news of another crisis. "I'm keeping the beach house."

Josh's face lit up and he stood up with his fists raised above his head. "Dude, yes!"

Josh fingered open the pouch and lifted the gold chain from the interior. At the end of the chain, two shards of the Bloodstone were configured in a cross held together by a golden thread at the crossing. Josh gasped. "This is part of the Bloodstone!"

"They dug those two pieces out of my chest. Any further and they would have pierced my heart. It's a present from me and, in a way, from your Uncle Cephas. I thought you would appreciate having the same stone that your father had."

Josh's eyes glittered and he dabbed at them as he put the necklace over his head. The small stone slid into his tank top and he placed a hand over it. "Thanks, Jonathan. It means a lot to me."

Steel reached over and punched Josh in the shoulder. It was all he could think of to do. "Just don't get any ideas about using it to augment your awesome mental powers."

"Hey, bro, my Dad did tell me I was a genius. Don't forget that."

"I won't." Steel said wistfully. They sat in silence for a long time and then Josh held up a finger.

"Think it will work on girls?"

"You need all the help you can get."

"Sweet!" Josh picked up the plate with the cake. "Dude, see those two girls walking down the beach?"

Steel glanced into the distance and watched two teenage girls playing in the water's edge. "Yes. You're out of their league, Josh."

Josh raised an eyebrow. "Not if I use the Bloodstone to summon them." He stood up with the cake plate in one hand and wiggled his hand in the air. "Let them eat cake. Let them eat cake." He chanted.

Something came over Steel. It was a distinctly unfamiliar sensation. Call it whimsy. Call it playfulness. Call it insanity. He reached up calmly and shoved the cake into Josh's face.

Josh froze and the cake and the plate fell away leaving his face covered with icing. It bubbled up inside of Steel. At first, it was painful like a hiccough. Then he laughed. And, he laughed some more. And, it began to feel good until he saw Josh move quickly. Before he knew it, Josh shoved Steel's piece of cake in his face. He stood up abruptly waiting for his anger to lash out. Josh's wide eyes glared back at him through the cake icing.

"Uh, oh. Bad idea." Josh ran down the stairs toward the beach and Steel chased after him. He caught him at the water's edge and a wave threw them both down on the sand. Steel sat up with seaweed hanging from his frosting covered face. Josh sat up beside him. Two shells were stuck on his cheeks in the cake frosting.

They looked at each other and Steel laughed. He laughed

and laughed until his injured chest hurt and tears ran through the cake frosting. Josh joined him until a shadow fell over them. They looked up into the faces of the two girls.

"Whose birthday?" The brunette asked.

Josh stood up clumsily and wiped the icing from his face. "Mine. I'm seventeen. Not sixteen. Seventeen. You know, one short of eighteen. Yeah, seventeen." He bit his upper lip and glanced at Steel in dismay. The other girl, a blonde reached out and wiped a small dab of icing from Josh's forehead. She popped it into her mouth.

"Tastes good. Aren't you going to invite us up for some cake?" She said.

Josh froze and Steel reached forward and shoved him in the back. "Josh, where are your manners? Take them up for cake."

Josh smiled and motioned toward the beach house. "Sure. Let's go." The girls ran toward the beach house and Josh turned back to Steel. He pointed to the Bloodstone under his wet tank top. "It worked." He ran off after the girls.

"Chief," Theo lumbered past Josh. He held Steel's cell phone in one hand and a towel in the other. "Don't you go wasting my good cake. Here, clean off your face."

Steel took the towel and wiped his face. "Sorry, Theo. I got carried away."

Theo nodded. "Need to do that more often." He held up the phone. "Call for you."

Steel handed Theo the towel. Theo hurried after Josh and the two girls. "Don't want leave that boy unchaperoned. You got a lot to learn about teenagers."

Steel pressed the phone against his ear. "Hello?"

"Steel, this is Miller."

Steel tensed. "What do you want?"

"Calm down. I'm not in the military anymore. And, I'm recovering nicely from my wounds suffered in the great battle against the Leviathan." He said.

"Yeah, thanks for helping Ross."

"Ross is not too happy. He's in a bit of hot water over the

whole operation. His partner, Sculder turned him in. The man is livid."

"What's new?"

"Anyway, like I said, I've left the military. There's nothing for me there anymore. Turns out Renee never changed her will after the divorce. I got my aviation company back. I'm going private. Although Renee almost spent the business into bankruptcy with all the global flights these past few days." He sighed. "You were Hot Steel, weren't you?"

"Yeah. I had a flashback. I was there that night. But, I came for another reason than to help you capture the children."

"You were after the Captain, weren't you? Well, I've decided to help you."

Steel blinked. "With what?"

"Finding the Captain." Miller said. "Once I go into the private sector, my sources will shut up. So, I did some last minute checking. Your father wasn't working alone when he went after the children. He was working for someone else. Someone bigger and more powerful. Someone with deep pockets and heavy duty political connections. My sources managed to pull out one number from all the loose ends. I'm texting it to you now."

Steel glanced at the screen as a series of numbers filled a text window. "What is this?"

"I think it's some kind of bank account number of the big cheese behind the Captain. A Swiss bank. You might want to start there. I'm out of the game. But, if you find out anything, I expect you to let me know. I have unfinished business with the Captain as well."

Steel nodded. "I will."

Miller was silent for a moment. "I understand you're having a birthday party. Don't worry. I'm not watching you. Theo mentioned it. Take care of Josh. Protect him. I will never get to know my own daughter."

Steel wanted to remind the man Vega wasn't his daughter, but he bit his lip. Josh wasn't his son, either. "I will."

The line went dead and before Steel could pull it away

from his ear, it rang again. He looked at the screen and three letters showed. "Max."

He recalled the day Josh left the courtroom in Dallas. He had made a call to 'Max'.

Jonathan pulled the pendant out of his pocket. On the side of the tiny key a series of numbers was etched so small he could barely see them. He picked up the pay phone and dialed in the long distance number followed by the string of numbers. He listened as the line clicked and whirred somewhere in Europe. Someone picked up the phone.

"Is she dead?" A voice with a foreign, mostly British accent asked.

"Max?"

"How did she die?" The voice continued calmly.

"Saving my adopted son." Steel whispered hoarsely.

Silence. For a moment Steel thought the man had hung up. "She redeemed herself?"

"Yes."

"Good. Call me back at this number in two months. I will be putting her affairs in order and then you can come and deal with it. I trust she made you promise?" The voice sounded as if it too were filled with emotion.

"Yes. And, I always keep my promises. Who are you?"

"Someone who tried to change Raven's heart. I'll speak to you in precisely two months from today." The line went dead.

Steel hung up the phone receiver and stared at it blankly. Who was Max? What function did he serve for Raven? And, should Jonathan be helping this man who could very likely have overseen the career of an assassin?

Steel answered the phone. "This is Jonathan Steel."

"That number just texted to you triggered an alarm, I am afraid. Now that you have it, we must talk." A woman's voice said. The last time they spoke, it was the voice of a man.

"How do I know you are Max? Your voice is different."

"I have my resources, Mr. Steel. I can sound like anyone but we both know Raven and what happened to her. Now, I know nothing about that number in the text. But, I need it as much as you do. By the way, I am sorrowful for your loss. Cephas Lawrence was a formidable man."

"How do you know this?" Steel tensed.

"I know things, Mr. Steel. That is my business. Do you still have Raven's amulet?"

"Of course. It's safe." Steel answered.

"Good. I assume that you want to honor Raven's request and help make reparations for all of her atrocities?"

"Yes."

"We have a deadline to observe. We cannot move on this until November. I cannot explain anymore on a potentially unsecure line, Mr. Steel."

November? That gave him a few months to recover. "What do I need to do?"

"I will send you travel information soon. It will probably be the week of Thanksgiving." Max said.

"Where am I going?"

"Switzerland." Max said. "I'll be in touch."

Steel drew a deep breath and turned to look around him. The clouds of a gathering storm brewed off over the ocean. Children played at the edge of the water. A wave had washed away the sand castle. Adults sat beneath umbrellas reading, napping, drinking, relaxing. Josh and the two girls were talking on the deck. Theo was serving more cake. Ordinary people in ordinary circumstances surrounded him. But, Steel now knew he was different from these people. He had a mission to fulfill. The admission of that fact carried with it a hint of his old anger and rebellion. But, now, the intensity had lessened. He had to accept the fact he was in this for a reason.

God had chosen him to battle evil. Pure and simple. Frodo and the ring. David and Goliath.

"I know you're listening." He spoke to the empty air. The Messengers were there on both sides of the battle; invisible, intangible, but real. "Vega called you invisible monsters. I don't care which side you are on. You manipulate us like we are puppets and that makes you more of a monster than I will ever be. Now, listen to me. I understand now I have a purpose. I know my mission. But, you will be there to help us. Do you understand?"

The wind ruffled his drying shirt and then died. He knew the angels had heard him. He glanced again at the cell phone with the cryptic number. And, he had no doubt that by November, he would be facing off against the next demon.

It was his destiny.

Epilog

Arthur Knight's lungs felt like they would burst. The cold, harsh sting of the vacuum of space sucked at his skin and pulled the life from him. Warmth played over him and his skin crawled with energy as the beam from the satellite engulfed him. His hair shriveled away even as his skin began to bubble and blister. Soon, he would be dead and all of this pain would end.

"You will not die this day, Arthur Knight." He heard the voice in his head. He tried to open his eyes but his lids were frozen to his corneas. Something warm and leathery wrapped itself around him. He blinked painfully and tore open his eyes. Donnatto had wrapped his wings around him and he could barely see the energy beam surrounding them both with sparkling waves of light. At the center of the beam something happened. It was a split in space; a tear in the fabric of other dimensions; a tiny crack in the edge of the universe. They were sucked into the tear and Arthur Knight was somewhere else.

Light and fabric and sound and silence all around him now and things reaching, touching, scratching at his scaly skin and there was Claire maybe somewhere smiling at him but why was she here and where was here and who are these black robed figures with steely masks and dark hoods and eyes of evil and malice; Penticle, Penticle, we are the Penticle awaiting the harbinger of the Redeemer; and more wings now of truly leather and black and spinning and spinning through a collage of colors; a curtain of cold rainbow energy so thick and imprenetrable and space winding down sunlight and spiraling nebulae and through metal to hard, cold floor.

Arthur gasped for breath and the pain was searing as he sucked air into his empty lungs. His skin burned with pain. Donnatto pressed his mouth close to Knight's ear.

"One moment and you will be healed." He whispered.

A sensation of warm fluid bathing him and the pain disappeared. His breath came easier. Donnatto released his wings and unclasped his arms and Arthur Knight was born anew falling from the embrace naked and sticky with the healing fluid onto the cold metal floor. Arthur blinked away the scales from his eyes and pushed himself up into a sitting position. The Bloodstone hung from around his neck, the only object to survive the trip to wherever they were. He looked up into the room in which he sat.

The walls were made of ancient stone. Glowing balls of green light floated in front of nine columns. He sat in the center of the room on a raised copper shield. Above him, a huge blue stone glowed with fading light. It was set into the dome of the room and each of the nine columns curved inward to end at the golden ring that held the huge jewel suspended above him. Donnatto stepped around him and surveyed the chamber. The messenger of God looked down at his arms and over his shoulder at his wings. He seemed to have shrunk and deflated, was the only word Knight could come up with. His wings no longer glowed with light.

"I no longer have access to the higher dimensions of my Master! I am now more like a mortal, Arthur Knight. This is the price I have paid to accompany you here." Donnatto's face was pale and he frowned.

"Where are we?" Knight slowly stood. A grating noise came from behind him and he whirled to watch as part of the stone wall slid aside. A stooped figure in a gray cloak shuffled in with his face hidden beneath a huge hood.

"We must hurry. With the Portal active again in so short a time, the Daemons will be here shortly. I hope you have brought the Harbinger of the Redeemer."

Knight gasped. He recognized the voice. "I know you."

Gnarled fingers appeared from the sleeves and grasped the hooded cloak. The hood slid back and Dr. Cephas Lawrence frowned. "Oh, my! It's you! Well, this is a most unwelcome surprise and if you do not want to die a heinous death, we must hide." He gestured to an arched window at the side of the chamber. In the distance, Arthur saw hideous winged figures soaring toward them.

"Where am I?"

Cephas Lawrence pushed a hidden switch and a secret doorway slid open. "Inside. Now. And, welcome to the Node of God."

Knight glanced at Donnatto. "The Node of God? Cobalt was right?"

Donnatto nodded. "And that means we must find the Children of Anak and Renee."

Knight caught one last glimpse of the approaching winged creatures through the window before the stone door closed around him. He almost wished he had died!

SOURCES:

"Lights in the Sky and Little Green Men"
"More Than a Theory"
"Why the Universe is the Way It Is"

And, now for your enjoyment and anticipation a part of the first chapter of "The 9th Demon: A Wicked Numinosity"

Chapter 1

Jonathan Steel felt the unmistakable jolt course through the underbelly of the airplane and he knew they were going down. Steel felt no fear, no sense of dread as the airplane's tossing and bucking worsened. He had faced death many times before. But, the people around him were not so lucky. An elderly man across from him was sweating and his face was pale. The woman next to him crossed herself and muttered silent prayers. In front of him, a mother held her toddler to her chest. The child's cries were lost in the roar of air outside the fuselage.

Steel's stomach lurched as the airplane wallowed in the sky. Objects floated in the air around him and he swallowed to keep his cola and pretzels in his stomach. People screamed and pandemonium ensued. The flight attendants frantically grabbed the seats as their feet lifted into the air and then slammed to the floor as the airplane regained power and climbed back into the night. Steel felt his head bounce against the seat in front of him and felt warmth trickle between his eyes. He touched the skin and his hand came away covered with blood. One flight attendant lay on the floor just feet away from him. Her leg was turned at an ugly angle and she was pinned to the floor by the airplane regaining altitude.

As in times before when he faced death, Steel waited for his life to flash before his eyes. It did not. His life before two and a half years before remained a mystery to him and he had recovered only bits and pieces of it. He took this as a positive sign. If his life did not flash before him, they he wasn't going to die.

The plane leveled off, still tossing in the turbulence and the flight attendant crawled down the aisle. Tears streamed down her face as she pulled the broken lower leg behind her. The air grew still and quiet. People fell silent. An oppressive sense of dread settled over the interior of the aircraft. Above it all as if the sound had focused only on him, Steel heard a woman crying.

Steel wiped the blood from his face and glanced back along the aisle for the source of the sobbing. The woman sat alone with her face etched in fear. She was studying something in her hands hidden by the seat in front of her. She raised it to the window of the exit row seat and a flash filled the darkness. Passengers around her jerked in shock. She had taken a picture through her window. The light from the camera's image illuminated her face. Her eyes filled with shock and she pressed a hand to her lips. Then, as eerie as it might have seemed in that strange setting, she looked up at Steel with familiar eyes. He blinked but could not recall where he had seen the woman. She motioned to him with her empty hand and pointed at the camera.

"You need to see this." She said. In the tense silence she might as well have shouted.

As if possessed by something beyond himself, Steel unbuckled his seat belt and stood up. A flight attendant barked at him from down the aisle and he ignored her. The floor shook beneath him and suddenly vibrated with the same deadening lurch he had felt before. He stumbled against a seat and whirled to plop into the empty seat beside the woman. Her eyes never left his and she handed him the camera. Dark hair cascaded around her face and she swallowed. She pointed to his seat belt.

"Put on your belt. Now that you are here, we will live." She whispered. Steel snapped the belt around his waist and examined the image on the camera. Before he could register his total and complete shock, the airplane spun in the turbulent air, turned upside down, and fell out of the sky.

Steel returned to consciousness to the grinding, harsh vibration of the airplane as it slid across snow. Cold air bathed his face and every muscle in his body ached. The plane had broken in half right through the row in which he had been seated. The front half of the plane was gone and only the black, frigid night filled the huge open front of the rear half of the airplane. Snow showered from the leading edge and filled the air with an icy storm. He tried to clear his clouded mind. They had been somewhere over the Alps on the way to Zurich, Switzerland when the turbulence had kicked in. They were sliding down the edge of a mountain, headed toward a collision with solid rock. He glanced at the woman beside him. Her head lolled against her chest but he didn't have time to see if she was breathing. He still held the camera with the impossible image from outside the window. He slid the camera's wristband over his right hand and reached across the woman for the emergency exit door. The only reason this portion of the airplane was still intact was because they sat over the wings. He strained and lifted the exit door out of the frame and more cold air blasted him from the side. He held onto the handle, fumbled with his seat belt and undid the flap.

Steel undid the woman's seat belt and slid over her still body, pulling her after him as he climbed out onto what remained of the wing. He looked once forward of the plane's path and saw the huge mountain cliff illuminated by scarce moonlight just a football field's length away. Wrapping his arms around the woman's body he crouched onto the exit door and slid off onto the snowy mountainside. Something huge lumbered over his head and he leaned away from the airplane's huge tail section. The exit door spun on the ice and he trailed his foot in the snow to stabilize their sliding.

Steel hit bump after bump and finally was thrown off of the sliding exit door. He wrapped the woman in his arms and thudded into the snow, feeling his ribs crack and his spine snap. It was the worst ski fall of his life. That is if he had ever fallen while skiing. He rolled and tumbled, holding the woman against him. He felt more than saw the collision of the rear section of the airplane with something huge and rocky down the mountain slope. The night grew bright with the explosion. Still he rolled, slid, and tumbled until his hands were raw and bleeding from the icy snow. His speed slowed and he came to a stop in a drift of hardened snow. The reflection of flames played across the rock strewn but snowy slope transforming it into an icy vision of hell.

Unable to move, Steel watched the plane as it was consumed by the fire. His mouth was filled with snow and blood. His ears were stuffed with the cold. He shook his head and only then heard the sound of dogs barking.

Was he hallucinating? He scanned the snowy slope back up the mountainside and saw nothing but rocks and snow. No dogs. No humans. He and the woman were the only survivors. He reached down and touched the woman's throat. Her pulse was strong. He knelt beside her and examined her limbs. No fractures he could see. Red drops splattered on her throat from the cut in his forehead. Already, his head was beginning to pulse with pain. He wiped the blood away and grabbed a handful of snow, pressing it against the cut on his head. It was only then he remembered the camera dangling from his wrist. Or, more appropriately, what was left of the camera dangled from his wrist. The lens was shattered along with the viewscreen on the back. He opened the tiny door and took out the memory card. He tucked it into his jean's pocket and tossed the broken camera away into the snow. Time to think about that later. For now, he could only think of the dead. So many dead. So many wasted lives. Across the snowy bowl in the mountainside, the flames of the burning airplane brought light to the darkness. He looked behind him down the slope. About a hundred meters away, he spied a pale, yellow light. He picked up the woman's limp body and a sharp pain ran through his right chest, probably a cracked rib. He stumbled through the deep snow toward the light. A long, low building appeared in the hellish shadows from the burning wreckage. He didn't understand the letters etched on the door in French. He kicked open the door and stepped into a relatively warm interior filled with the odor of dogs. Twelve dogs in kennels were lined up against the far wall. They barked and howled in fear and pain as the sound of the burning aircraft crackled through the night. What were dogs doing here on the mountainside?

A gas stove cast a pale, blue light and he laid the woman on a couch. He found the thermostat on the stove and turned up the heat. Dizziness swept over him and he stumbled back onto a chair next to the couch. The woman moaned beside him.

"Where am I?"

"In a hut on the mountainside." Steel turned to study the woman's open eyes. The bright green shade of her eyes gleamed in the stove light as she sat up beside him. "With dogs."

"You saved me." She pushed her dark hair back away from her face. "I knew you would save me."

"You saved me, it would seem." Steel blinked as the dizziness worsened. "If I had stayed in my seat, I would be dead right now."

The woman stood up and crossed to the window in the door to the hut. She wrapped her arms around her as flames from the distant airplane wreckage cast flickering light across her features. "Is anyone else alive?"

"I doubt it. The airplane broke in half and both sections are in flames. I didn't see any survivors on the mountainside." He felt nauseous and blood ran from his forehead again.

The woman turned her back on the door and studied him. "You are a kind man. Why did you help me?"

"It's what I do. I help people." Steel mumbled. He spied bottled water beside the chair and reached for one. He twisted the top off and guzzled the cold liquid. "I don't expect you to understand."

"I don't. But, it makes a difference. I think I know where we are. Jungfreud. If we are at the top, then this is the dog sled building. They let people ride on a dog sled when they get here from down in the valley." She crossed the room and reached for a warm one piece overall hanging on the wall. "I'm cold. I'm going to put this on."

Steel watched her shrug her way into the overalls. He tried to stand but the dizziness made movement impossible. "I'm very dizzy right now."

"Probably a concussion." The woman said. She opened a door in the rear of the hut and smiled. "Good. The sled is still here."

Steel tried to stand again and fell back into the chair. His body began to shake with rigors not so much from the cold as from the shock he had just been through. "I don't know if I can make it."

The woman squatted before him. She reached into the overalls and pulled out a small, cloth purse. "My dear sweet man, you are indeed fortunate. Since you did save my life, I will not take yours."

Steel felt a new chill run down his spine. And, it was not from the cold. "What?"

The woman retrieved a plastic packet from within the purse containing white powder. "I will take the dog sled down the mountain. But first, I will take your memory."

Before Steel could move, the woman poured a small measure of the powder into the palm of her hand and blew it into his face. He inhaled the acrid particles and his mind grew cloudy, muddled. Faintly, he heard the dog's barking grow louder and he pulled himself up out of the chair, lurching toward the doorway. He never made it.

CPSIA information can be obtained
at www.ICGtesting.com
Printed in the USA
LVHW110358261219
641704LV00001B/7/P